NSTA PATHWAYS
To the Science Standards

High School Edition

Editors
Juliana Texley
Ann Wild

NSTA Pathways to the Science Standards:
Guidelines for Moving the Vision into Practice

HIGH SCHOOL EDITION

Editors: Juliana Texley and Ann Wild
NSTA Project Coordinator: Sheila Marshall

Contributors: Hans Andersen, James Biehle, Patricia Bowers, Jeffery Callister, Michael Clough, Janet Gerking, Carole Goshorn, Jennifer Grogg, Mary Gromko, David Kennedy, Terry Kwan, Norman Lederman, Maria Lopez-Freeman, Joseph A. Miller, LaMoine Motz, Karen Ostlund, Michael Padilla, Larry Read, Patricia Rourke, Len Sharp, Victor Showalter, Gerald Skoog, Dennis Smith, John Staver, Sharon Stroud, Anne Tweed, Sandra West, Rachel Wood, and the NSTA Committee on High School Science Teaching.

NSTA is grateful for the generous contributions by the funders that made this effort possible:

Monsanto Fund
DuPont
American Petroleum Institute
Chemical Manufacturers Association

NSTA would like to thank the National Research Council of the National Academy of Sciences for permission to reprint material that originally appeared in the National Science Education Standards (Washington, DC: National Academy Press, © 1996)

Gerald Wheeler, **NSTA Executive Director**
Phyllis Marcuccio, **NSTA Associate Executive Director**
Marily DeWall, **NSTA Associate Executive Director**

Book Design: Elinor Allen, Art Director, Allen and Associates, Ltd.
Washington, D.C.

Watercolor Illustrations: Elizabeth Allyn

Library of Congress Catalog Card Number
Stock Number: PB 126X
ISBN Number: 0-87355-144-3

Third Printing, 1998
Copyright ©1996 NSTA
National Science Teachers Association
1840 Wilson Boulevard
Arlington, Virginia 22201

CONTENTS

Acknowledgements
1   Introduction
4   How To Use This Book

Science Teaching Standards

7   Realizing the Teaching Standards

Professional Development Standards

26   Reaching for the Professional Development Standards

Assessment Standards

41   Exploring the Assessment Standards

Content Standards

62   Mapping the Content Standards
65   Physical Science
87   Life Science
109   Earth and Space Science

Program Standards

127   Moving into the Program Standards

System Standards

149   Navigating the Systems Standards

Appendices

170   Appendix A: National Science Education Standards: Some Relevant History
173   Appendix B: National Science Education Standards
181   Appendix C: Designing High School Science Facilities
191   Appendix D: Resources for the Road: The CD-ROM Collection

Introduction

It is rare when a profession achieves national consensus in its vision for change. The National Science Education Standards represent just such a landmark effort. Science educators, scientists, administrators, businesspersons, and concerned citizens, coming from very different perspectives, have joined forces to speak eloquently for science education in our schools.

A document of this scope must offer a broad and all-encompassing vision. The Standards present a goal for science education—a destination that may seem distant to busy teachers in schools across the United States. In classrooms where the staff has taken on many social responsibilities apart from education, science teachers may find it difficult to read the Standards and determine where to begin to implement them. The very breadth of this landmark document may seem at first intimidating—something for everyone—but nothing for tomorrow's lesson plan.

So, for science teachers across America, we offer this practical guidebook, *NSTA Pathways to the Science Standards: Guidelines from Moving the Vision into Practice*. In its pages, we demonstrate how you can carry the vision of the Standards—for teaching, professional development, assessment, content, program, and system—into the real world of your classroom and school. This book is also a tool for you to use in collaborating with the administrators, school boards, and other stakeholders in science education.

Pathways was created by an impressive partnership of teachers and administrators who believe that many elements of the Standards are al-

Principles Underlying the National Science Education Standards

- Science is for all students.
- Learning science is an active process.
- School science reflects the intellectual and cultural traditions that characterize the practice of contemporary science.
- Improving science education is part of systemic education reform.

Reprinted with permission from the *National Science Education Standards.* © 1996 National Academy of Sciences. Courtesy of the National Academy Press, Washington, D.C.

ready in place today. Great science education is common, if not uniform, and today's teachers *can* achieve the Standards in the coming decades.

Many Options, No One Way

In schools nationwide, the Standards and the *NSTA Pathways* travel guide to the Standards will spark changes that will spring from the strengths of each school faculty and the unique character of each community. We stress that there is *no single, correct pathway* for improvement—no linear progression from "bad" to "good" in attempting to move the vision of the Standards into reality.

There are as many paths to change as there are modes of learning in our students. We can begin by building on what works well in our classrooms and schools today. We do not have to reinvent the wheel. Pilot projects and established programs that illustrate the vision of excellence can be found in classrooms across the country. Each teacher and each school is unique, and each journey to better science education will follow a slightly different path.

All *Students Can Learn Science*

All students can learn science, and *all* students should have the opportunity to attain high levels of what is defined as scientific literacy. This is one of the strongest of the four foundational themes (previous page) of the National Science Education Standards. In the examples in this text, you will find approaches that encompass different learning styles, different levels of reasoning ability, and different cultures.

The Standards draw strength from the diversity of applications that can be derived from them. Implied in the principle of "science for all," for example, is the recognition that some compensatory education may be needed to support learning for the science-shy or school-phobic, particularly in the secondary years.

It is clear, too, that tracking secondary classes will no longer be best practice in science education. Providing paper-and-pencil curricula to the students who "shouldn't be trusted" with the glassware is not in the best interests of students, schools, or American society.

The content delineated in the Standards is a goal for *all students,* not a layer cake with some content appropriate only for some students and some teachers. Naturally, the depth of knowledge will be modified for students with various career goals. But the essential experiences of inquiry, exploration, and application must be provided to *every* science student in the nation.

To do this, we will need new partnerships. Co-teaching science with specialists in remedial language or mathematics becomes more common every year. Bilingual classes provide quality science experiences to those for whom the language is a barrier. Most important, *smaller* classes must be the norm, so that the attitudes and behaviors required for hands-on science can be reinforced in a class of diverse learners in a safe, orderly environment.

Learning Is an Active Process

Another strong theme in the Science Standards is that learning science is an active process. The Standards rest solidly on the foundation of educational research that demonstrates that learning is an active process achieved by enthusiastic students. Much of the literature that supports this perspective uses the term "constructivism" to represent the student's role in building concepts.

This perspective does not diminish the importance of scientific knowledge. There is an *unshakable* base of content in the Standards. Becoming involved in the process of science is necessary, but not sufficient, for the learner. The Standards assert that science is learned *in* the content, but

is *not limited to* the content. Students must build their own concepts, but they must also continually compare and correct them by exploring the natural world and the body of knowledge that scientists have amassed.

For some, active learning may imply a dramatic change in the way we structure courses, programs, and systems. For example, if students are to be given the opportunity to compare their preconceptions to their experiences, far more time will be necessary in the content areas. The well-established methods of inquiry are not only desirable but also absolutely necessary for students to construct ideas, test them, and, if necessary, reject them and begin again in their search for ideas that more accurately reflect the real world. This, too, requires time.

Beginning the Journey

The primary strength of the National Science Education Standards is the process through which its vision was created: a consensus of divergent viewpoints forged of shared commitment. The document paints a clear picture of what should be taught (content standards), how to do it (teaching and program standards), how you will know when you get there (assessment standards), and how to build capacity for change (professional development and education system standards).

We invite you to join the partnership and follow your own pathway to the vision set forth in the Standards.

How To Use This Book

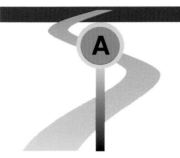

In each chapter, sign posts guide you on pathways through discussions about the Standards.

Because of their scope, the National Science Education Standards must speak to a wide variety of audiences: administrators, teachers, students, parents, legislators, businesspersons, scientists, social activists, and communities. All of them have a stake in science education, and all must play a role in improving America's scientific literacy.

The *NSTA Pathways to the Science Standards* has *one* audience—you, the high school science teacher. *Pathways* is *your* guide to the Standards. *Pathways* includes tools to help you move your teaching, professional development, assessment, program and curriculum, and interactions with the education system toward the vision of the National Science Education Standards. We intend this book to be practical and user-friendly.

The first three and last two chapters discuss the Standards that apply to all K–12 teachers: teaching, professional development, assessment, program, and system. While the messages in these chapters parallel those of the Standards, you will find suggestions specific to the high school teacher.

In each chapter you will also find Resources for the Road—a list of pertinent articles, most of which you can access on the compact disc that is available from NSTA as a supplement to this book. Each chapter also contains a list of the Standards and a chart highlighting the shifts in emphasis envisioned by the Standards. (Both of these are reprinted from the *National Science Education Standards.)* From the practical discussions about each of the standards, we trust you will find a few good pathways to travel.

The fourth chapter is devoted to the science content outlined in the Standards for students in grades 9–12. The casual reader may find the content daunting; in fact, it is a far slimmer core of knowledge than would previously have been required in most secondary courses. "Learn the core ideas, and learn them well" is the message of the Standards.

An important challenge in preparing the content chapter of *NSTA Pathways* was how close to stay to the organization of the Standards. The Standards represent the well-honed consensus of many professionals. To those concise statements of belief and practice, the *NSTA Pathways* remains scrupulously true.

In this book, however, we made one organizational change to match the current subject-oriented identity of secondary teachers. We hope this allows you to locate specific information quickly.

We have clustered the Content Standards in three sections: Physical Science, Life Science, and Earth and Space Science. For each discipline we include examples of Inquiry, Science and Technology, Personal and Social Perspectives, and History and Nature of Science, integrating them within the disciplines

Resources for the Road cited throughout the book lead to a CD-ROM collection available from NSTA.

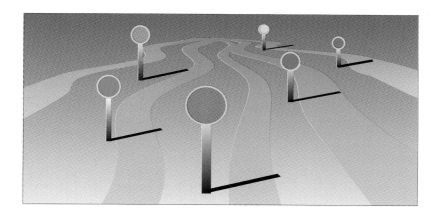

rather than devoting separate sections to them. This is the approach, we believe, that will be taken in most secondary schools.

We've included vignettes from exemplary programs and resources for more information. No single example stands alone. We've also suggested how you can make strides toward integrating your program.

No single teacher and no one class or school can move in all directions at once. Site-based planning teams may find some ideas in the *Pathways* already in place and others far beyond the reach of their community. For each situation, we trust you will find a few useful pathways to choose.

This is not an easy journey. While this text describes pathways, it is not a collection of recipes. We haven't provided lesson plans to copy for tomorrow's class or photocopyable masters to duplicate a lab exercise. Achieving the Standards cannot be prescriptive. The changes that will occur in *your* classroom will be unique and appropriate for you and for *your* school, *your* district, and *your* community.

We hope that *Pathways* will be a catalyst for local brainstorming and a readable source of inspiration. The text is deliberately concise to give you information that you can easily share with policymakers in your school, district, and state. We encourage you to pass this information on to principals, administrators, boards of education, parents, business representatives, legislators, and other citizens to help them understand what is needed for tomorrow's classrooms. We hope that over time you will continue the dialog sparked by the release of the Standards and that *you* will become a leader for science education in your community.

But we cannot undertake this effort alone. The *National Science Education Standards* is clear on this point:

"[I]t would a massive injustice and complete misunderstanding of the Standards if science teachers were left with the full responsibility for implementation. All of the science education community—curriculum developers, superintendents, supervisors, policymakers, assessment specialists, scientists, teacher educators—must act to make the vision of these standards a reality."

Both the *Standards* and *NSTA Pathways* present choices you and your school can implement now. We are not describing the perfect classroom or the ultimate teacher. As our shared journey progresses, our images will change and mature—just as the content and processes we call "science" change as we head into the 21st century.

Come with us as we journey together. The Standards offer the most exciting vision produced by our profession in this century. Take this *NSTA Pathways* travel guide along with you, and share the wisdom and guidance that our profession offers.

Science Teaching Standards

As teachers, we are the most important force in the education system. We are the key to reform.

Realizing the Teaching Standards

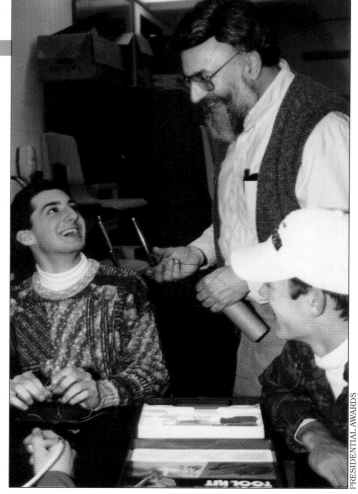

As teachers, we are the most important force in the education system. We are the key to reform. In each of our classrooms, we are the ones who will define the first steps on every pathway toward the Standards. Without the commitment of teachers, little progress can be made.

But to move beyond the good intent of individual teachers to national reform, we will have to share a much stronger common sense of direction than we have ever had before. And we need the support of *all* the other stakeholders in the educational system.

Although teaching may seem deceptively simple to the public, we know that what happens in our classrooms every day is a professional, extremely complex, and very personal activity. Our own views about the nature of science and learning—formed by how *we* were taught—affects everything we do. Changing theories and societal pressures can pass by our classroom doors with little effect if we have been so molded by our own educational history that we find it hard to change.

The five Teaching Standards challenge us to grow in each of the following areas:
- planning a science program
- guiding and facilitating learning
- assessing, learning, and teaching
- designing and managing the physical environment
- building learning communities
- school planning

Where Are We Going?

In Wonderland, Alice asks which way to go. She is told, "That depends a good deal on where you want to get to." To set out on a pathway toward the Standards, our profession first had to define its goal: Are we preparing students for college, for the workplace, or for personal fulfillment in adult life? The Standards answer this question: We teach *all* students science so that they will be able

- to use scientific principles and processes appropriately in making personal decisions
- to experience the richness and excitement of knowing about and understanding the natural world
- to increase their economic productivity
- to engage intelligently in public discourse and debate about matters of scientific and technological concern

So the destination is clear: It's not college, but adult life. That answer is one we can readily accept. But if we are to make long-lasting changes in that direction, not only teachers, but also principals, parents, and other community members will have to agree that college entrance should no longer be the most important measure of success.

The Standards challenge our preconceptions about what we do in our classrooms and encourage us to become more effective. As David Berliner summarizes in his essay, "Knowledge Is Power," "[W]e are on the threshold of creating a scientific basis for the art of teaching which will be acceptable to the general public as truly specialized knowledge."

We must enrich our vision of "science as an enterprise" and "science as a subject to be taught." We must convey what methods work and why, and we must realize that we already have many of the skills we will need to move forward with confidence in putting the vision of the Standards into practice.

How To Facilitate Learning

A central theme in learning theory today is a perspective called constructivism. This body of research is not new; rather it is a coherent integration of many ideas veteran teachers recognize. If John Dewey, Jean Piaget, and Benjamin Bloom could join us in a roundtable on constructivism, they would certainly find more common ground than disagreement. What's different today is that our research data is stronger, our conclusions more compelling, and our understanding of cognition much better than in the past.

For years, we have been frustrated by the apparent contradictions in learning research. Results have shown that very different methods of teaching produce similar gains in learning. Berliner's "science of teaching" and the unifying principle of constructivism now explain why this is so. We now know that we cannot just transfer, or hand over, our own understandings of the natural world to our students. Instead, as the Standards point out, students have to construct or build their *own* knowledge in "a process that is individual and social."

If we move past the jargon and statistics, the essential message of constructivism is at once revolutionary and simple: What most people consider "teaching" is impossible. There is only learning and facilitating learning. Constructivism does not deny that objective content knowledge must be learned, but it does imply that it may take some students longer to reconcile their preconceptions to their experiences than we might previously have believed.

Our role as teachers is to help students enthusiastically build accurate concepts and reject incorrect ideas along the way. We can engage students, structure time, create a setting, make tools available, identify resources, assess students' progress, and guide students' self-assessment. What we *cannot* do is build a learner's personal view of the world for her or him—just as the education system cannot build our own personal view of instruction by issuing directives. Many methods can prove successful, and many pathways can accomplish goals.

Kenneth Tobin, a prolific writer in the learning field, describes the impact of constructivist research as "paradigm wars." The battles being fought are less about *what* we must change and more about *why* we must change. And trying to give successful teachers a very different perspective on what they do may well be the most difficult task at hand.

There are many tensions in trying to teach toward the Standards: tensions between depth of understanding and breadth of content coverage, between the tendency toward teacher-directed activities and the desire to have students set their own goals, and between structures that work for many students and modifications necessary for diverse learners. Teaching is not simply a matter of style but a process of rational decisionmaking, carried out hundreds of times every day about each of these issues.

Doing Things Differently

How can we choose better paths? Research can now help us distinguish *what* we do from *why* we do it. Most of us know that we have to engage students' interest, allow them to explore, and encourage them to question incessantly—even if our main goal is only to motivate them toward rote learning. Most of us admit that lecturing doesn't work, even though we may not have the skills or the facilities and the support to abandon this approach.

So what at first seems to be a revolution may really be just an evolution. Most of us already know how to do great demonstrations, and we also realize that we do them not as "show and tell" but to help students confront their own misconceptions. Many of us already value questioning and need only a little encouragement and guidance to become better questioners. Stated simply, the step from "recipe" laboratories to true inquiry involves *who* will generate the problem to be explored.

As we move toward the Standards, our classrooms will begin to look and feel different. The doors will be open more often, with teams of teachers planning together for long-term goals. Textbooks will be slimmer, thanks to honed-down content, and will be only one resource among many. The range of resources that students will use to reach their content goals will be far broader than ever before. Science students will access the resources of the community and the world as they work to construct knowledge.

In tomorrow's classroom, we as teachers will continually assess what students are learning. The pace, structure, and level of instruction will vary for each student and for each topic. Tomorrow's classroom will be a more social place, since teachers will recognize the social component of learning. Collaboration and discourse will be constant components of lessons, as we assume the role of lead student in a community of learners.

In the Standards, "inquiry" is the term used for the content knowledge of how to do science. We support students in learning how to answer their own questions through media research and scientific investigation. Inquiry is also a method discussed in the Teaching Standards.

Planning, guiding, and assessing learning are three other aspects of teaching. Designing the classroom environment, building communities of learners, and planning school-based programs will extend our journey beyond supporting individual learners to assuming our roles in the larger school system. To each aspect we can bring a wealth of shared experience.

The sections that follow explore each of the six Teaching Standards to help you begin your journey. If the journey of 1,000 days begins with a single step, what are we waiting for?

Resource for the Road

Berliner, David C. (1987). In David C. Berliner and Barak V. Rosenshine (Eds.), *Talks to Teachers*. New York: Random House.

About Teaching Standard A

Planning a Science Program

Teachers build on their thorough understanding of science content to plan effective learning environments for students. While long-term goals may be keyed to district and national outcomes, the task of daily planning involves great sensitivity to varied learning styles, abilities, experiences, and motivations.

Pacing Science Learning

Every hour of every school day, we make hundreds of decisions about student learning. Engage, assess, explain, review, explore, extend, motivate, move on... The minute-to-minute pacing of instruction is the heart of teaching. In constructivist terms, as students build their own understandings of the world around them, good teachers stand at their side, handing them the right tools and putting the raw materials, broken down into manageable units, within reach.

Programs and systems establish outcomes of education, destinations for students and teachers to reach. But only teachers can determine the pace students will take on that road. The National Standards provide clear direction for the journey.

The changes required are subtle but dramatic. Where traditional programs often relied on content to determine the direction of instruction and the speed at which the class progressed, a Standards-based program will be paced by the learner. Where teachers (for whom the subject comes easily) might take a straight path toward a content goal, a learner-directed curriculum will take what might seem at first glance to be an indirect or even circuitous route.

While teachers may collaborate to establish the year-end outcomes of courses of study, the day-to-day goals that they set for their students must, of necessity, always be unique. Each unit must begin with students exploring their own preconceptions, or naive (untutored) models of reality. Learning starts with students and their own experiences.

The process may be different in different classrooms: brainstorming, demonstrating and discussing, drawing, journal writing, or predicting. It doesn't matter *how* we assess where to begin, but it is absolutely essential *that* we do. Research shows that unless students acknowledge their naive ideas about the

TEACHING STANDARD A

Teachers of science plan an inquiry-based science program for their students. In doing this, teachers

- develop a framework of year-long and short-term goals for students
- select science content and adapt and design curricula to meet the interests, knowledge, understanding, abilities, and experiences of students
- select teaching and assessment strategies that support the development of student understanding and nurture a community of science learners
- work together as colleagues within and across disciplines and grade levels

Reprinted with permission from the *National Science Education Standards.* © 1996 National Academy of Sciences. Courtesy of the National Academy Press, Washington, D.C.

10   Realizing the Teaching Standards

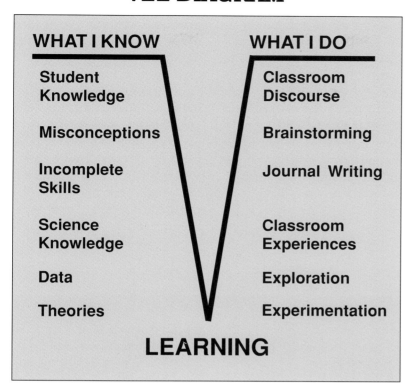

VEE DIAGRAM

WHAT I KNOW
- Student Knowledge
- Misconceptions
- Incomplete Skills
- Science Knowledge
- Data
- Theories

WHAT I DO
- Classroom Discourse
- Brainstorming
- Journal Writing
- Classroom Experiences
- Exploration
- Experimentation

LEARNING

ponents of a concept and the relationships between and among them.

Since each learner sees the domain of knowledge in a slightly different way, the teacher-drawn concept map will always be somewhat personal. Many of us find that having students develop concept maps is not only a valuable learning exercise but also an effective assessment of understanding.

While concept maps encourage both divergent thinking and the making of connections, Gowin's Vee diagrams, or maps (left), help teachers and students focus thinking into succinct questions to be answered. The left side of a concepts they study, they never really abandon them.

Once we know where most students' own paths to learning in a particular area begin, we can proceed to diagram or map the way toward student understanding.

There are many valid methods for planning learning, but two—concept maps and Vee diagrams—currently dominate the literature in science education.

The concept map (right) is a tool to organize and represent knowledge. Even a single lecture or lesson might involve 30 or 40 separate but related concepts. The map allows us to analyze the com-

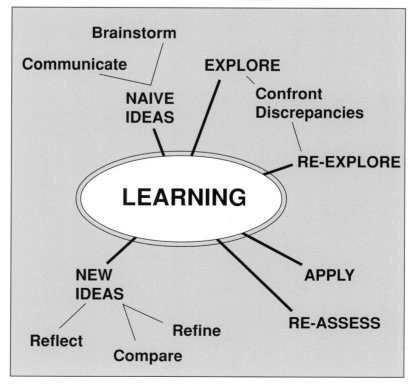

CONCEPT MAP

Centered on LEARNING with connections to: Brainstorm, Communicate, NAIVE IDEAS, EXPLORE, Confront Discrepancies, RE-EXPLORE, APPLY, RE-ASSESS, Refine, Compare, NEW IDEAS, Reflect.

Realizing the Teaching Standards 11

Vee map lists what the mapper knows; the right side shows what is being done and can be done. A Vee map focuses on how thinking and doing interact in trying to solve problems or answer questions.

Whether we use concept maps, Vee diagrams, or our own original representations, the task is the same. To create effective learning environments, we need to develop invisible antennae to help us locate the various components that contribute to understanding a concept and the individual perspectives that students bring to understanding a particular concept.

To reach the vision of the Standards, we will need to hone our skills in planning, designing programs, and in individualizing and modifying class programs for every learner.

We will also have to demand and be granted the necessary time. It is clear that pre-assessment, classroom discourse, and continual feedback all take far more time than we have now. Given enough time and opportunity to create effective learning environments, we can nurture learning that lasts beyond the next multiple-choice test. Without that time, "top-down learning" with its all-too-familiar short-term results—memorized today, forgotten tomorrow—will continue.

Resources for the Road

Association for Supervision and Curriculum Development (ASCD). (1994, February). Teaching for Understanding. (Theme Issue). *Educational Leadership, 51* (5).

Hausfather, Samuel J. (1992, November/December). It's Time for Conceptual Change: A Flexible Approach Leads to Understanding. *Science and Children, 30* (3), 22–23.

Johnson, Janice K. (1991, October). On Research: Focus on Pre-reading Activities To Improve Student Reading Comprehension. *Science Scope, 15* (2), 32–34.

Koch, Janice. (1993, March). Face to Face with Science Misconceptions. *Science and Children, 30* (3), 39–40.

Novak, Joseph D. (1991, October). Clarify with Concept Maps. *The Science Teacher, 58* (7), 45–49.

Novak, Joseph D. (1993, March). How Do We Learn Our Lesson? *The Science Teacher, 60* (3), 50–55.

Roth, Wolff-Michael. (1990, April). Map Your Way to a Better Lab. *The Science Teacher, 57* (4), 31–34.

Roth, Wolff-Michael. (1993, February). The Unfolding Vee. *Science Scope, 16* (5), 28–32.

Roth, Wolff-Michael, and Bowen, Michael. (1993, January). Maps for More Meaningful Learning. *Science Scope, 16* (4), 24–25.

Roth, Wolff-Michael, and Verechaka, Guennadi. (1993, January). Plotting a Course with Vee Maps. *Science and Children, 30* (4), 24–27.

Schamp, Jr., Homer W. (1990, December). Model Misunderstandings. *The Science Teacher, 57* (9), 16–19.

Stepans, Joseph, and Veath, M. Lois. (1994, May). How Do Students Really Explain Changes in Matter? *Science Scope, 17* (8), 31–35.

Tobin, Kenneth (Ed.). (1993). *The Practice of Constructivism in Science Education.* Washington, DC: American Association for the Advancement of Science (AAAS).

About Teaching Standard B

Guiding and Facilitating Learning

To guide students in learning, teachers must understand what learning entails. New research suggests that learning is an active process. Each student is responsible for her or his own learning.

Constructivism in the Classroom

While it's been called the most exciting idea of the past 50 years, the learning theory called constructivism has also been accused of turning education upside-down. Ernst von Glaserfeld, whose work has led the field, challenges the most basic assumption that guides teaching: that knowledge can be communicated through language.

Backed by a rapidly expanding body of research, constructivists contend that typical schooling cannot permanently alter the naive misconceptions that students bring with them to school and that standardized tests tell us little about what students understand and can apply. To quote Hans Anderson's oddly accurate grammatical error, "You can't learn anybody anything!" Put another way, students have to learn it on their own.

A constructivist approach does not imply that every student must rediscover every scientific fact and theory. Rather, as each concept is presented, a student compares it to the internal structures (preconceptions) that are already present. If the new information or experiences fit, they are incorporated into the learner's view of the world. If they don't, there must be time in the curriculum to reconsider what had been assumed and to rebuild the structures.

Despite its revolutionary conclusions, constructivist research offers a wealth of applications that have been clearly demonstrated to increase learning in school settings. Among the applications are these:

Begin by Assessing Student Preconceptions. Until you (and your students) realize what they think they know about a topic, building new ideas will be difficult. Interviews, questions, drawings, predictions, group discussions, and even more thorough pre-testing can help to uncover preconceptions.

Build on past experience. Meaning is created in a student's mind when concrete, physical experiences

TEACHING STANDARD B

Teachers of science guide and facilitate learning. In doing this, teachers

- focus and support inquiries while interacting with students
- orchestrate discourse among students about scientific ideas
- challenge students to accept and share responsibility for their own learning
- recognize and respond to student diversity and encourage all students to participate fully in science learning
- encourage and model the skills of scientific inquiry, as well as the curiosity, openness to new ideas and data, and skepticism that characterize science

Reprinted with permission from the *National Science Education Standards.* © 1996 National Academy of Sciences. Courtesy of the National Academy Press, Washington, D.C.

interact with existing beliefs and understandings. As long as these experiences fit existing frameworks (called schema), nothing will change. But when an experience contradicts prior knowledge, the learner becomes surprised, frustrated—and then, perhaps, willing to learn.

Take Sufficient Time. The literature is clear that discarding one's previous views of the world does not come easily. If it's convenient, many students will simply opt to ignore new information, especially if it contradicts previously held beliefs. Restructuring beliefs about the natural world and constructing new knowledge will only happen if a student is given sufficient experience, time, and motivation to make it happen.

Foster Continuing Inquiry. Through questioning, self-assessment, and redesign of traditional hands-on experiences to open-ended ones, constructivist teachers must constantly arrange learning environments that challenge students to create more accurate knowledge for themselves.

Constructivist research may be disconcerting to traditional lecture-oriented education programs, but it presents strong empirical data for what we regard as "best practice" in science education.

For example, direct sensory experiences (such as lab investigations) are crucial in encouraging students to question their preconceptions. Such time-honored methods as "thinking out loud," developing hypotheses, interpreting data, offering constructive arguments, and making predictions are part of a constructivist classroom. Small-group work stimulates higher-order questioning and effective learning.

Constructivism challenges us to give up many methods we've held on to far too long: almost useless lectures, cookbook labs (which should be replaced by open-ended inquiry), and memory-oriented objective tests that tell us little about student growth.

Is constructivism just another fad in the cycle of educational methods? Paul Brandwein and Lynn Glass (1991) suggest that it is really part of a "permanent agenda." The very process of scientific research supports a constructivist approach in classrooms and thus a clear path to the Standards.

Resources for the Road

Andersen, Hans O. (1991, May). Y'all Can. *Science Scope, 14* (8), 28–31.

Brandwein, Paul, and Glass, Lynn. (1991, April). A Permanent Agenda for Science, Part II: What Is Good Science Teaching? *The Science Teacher, 58* (5), 36–39.

Canady, Robert Lynn, and Retig, Michael G. (1995, November). The Power of Innovative Scheduling. *Educational Leadership, 53* (3), 4–10.

Cherif, Abour. (1993, December). Relevant Inquiry: Six Questions To Guide Your Students. *The Science Teacher, 60* (9), 26–28.

Clough, Michael P., and Clark, Robert. (1994, February). Cookbooks and Constructivism: A Better Approach to Laboratory Activities. *The Science Teacher, 61* (2), 34–37.

Hackmann, Donald G. (1995, November). Ten Guidelines for Implementing Block Scheduling. *Educational Leadership, 53* (3), 24–27.

Rezba, Richard J., Cothron, Julia H., and Giese, Ronald N. (1992, February). Traditional Labs + New Questions = Improved Student Performance. *Science Scope, 15* (5), 39–44.

Roth, Wolff-Michael. (1991, April). Open-Ended Inquiry: How To Beat the Cookbook Blahs. *The Science Teacher, 58* (4), 40–47.

Saunders, Walter L. (1992, March). The Constructivist Perspective: Implications and Teaching Strategies for Science. *School Science and Mathematics, 92* (3), 136–141.

Tobin, Kenneth (Ed.). (1993). *The Practice of Constructivism in Science Education.* Washington, DC: American Association for the Advancement of Science (AAAS).

About Teaching Standard C

Assessing, Learning, and Teaching

To pace the instruction in a Standards-based classroom, teachers must constantly gather data on their students. The easiest and most effective source of student data is the question. Guide your instruction with the right questions, and soon your students will be assessing themselves the same way.

The Questioning Classroom

A recent article with the title "How To Annoy Students" answered the dilemma simply: Ask them questions! Good questions *are* annoying—*and* stimulating *and* exciting *and* indispensable if you want to pace instruction through continual assessment.

But while most of us form questions by intuition (Paul Otto, 1991), with a little practice we can raise the level of questioning and, in turn, the amount of assessment in our classrooms (Gilbert, 1992).

The most common kind of classroom question, quick recall, elicits the lowest level of response. Examples are "What is the definition of mass?" and "When was the electric light bulb invented?"

Faced with such questions, students can either be "right" or "wrong," and often the quick answer is out of reach. Of course, if your goal at the moment is at the "knowledge" level, such questions can gauge the progress of your class.

But it's easy to move up the classroom taxonomy and collect better information about how your students are constructing knowledge. Consider these examples offered by Gilbert:

Comprehension Questions
   "What can change this result?"
   "What will happen next?"
Application Questions
   "What is the force on object C?"
   "What will the effect of treatment be?"
Analysis Questions
   "Which statements are inferences?"
   "What is the connection between drugs and crime?"
Synthesis Questions
   "Based on the data, what is your conclusion?"
   "How would you test your hypothesis?"
Evaluation Questions
   "How is your position

TEACHING STANDARD C

Teachers of science engage in ongoing assessment of their teaching and of student learning. In doing this, teachers

- use multiple methods and systematically gather data about student understanding and ability
- analyze assessment data to guide teaching
- guide students in self-assessment
- use student data, observations of teaching, and interactions with colleagues to reflect on and improve teaching practice
- use student data, observations of teaching, and interactions with colleagues to report student achievement and opportunities to learn to students, teachers, parents, policymakers, and the general public

Reprinted with permission from the *National Science Education Standards.* © 1996 National Academy of Sciences. Courtesy of the National Academy Press, Washington, D.C.

consistent with hers?" "In what way is your method scientific?"

To pace a class with questioning, we need to slow the rate of conversation markedly. We must encourage students, too, to be patient and discover how to become involved in the questioning process. They must learn to sit quietly and give the questioned student time to reflect. They must also learn when to support another student with a suggestion or an answer.

As teachers, we must master the art of reacting to questions. Otto suggests that we learn when to accept or reject, when to elicit clarification, when to ask another question, and when to move on.

We can learn a lot about students through the answers they give, but we can learn even more from the questions they raise. According to Fred Vincent (1993), "what if..." questions can be infectious in a classroom. Janet Chahrour (1994) takes questioning a step further, asking groups of students to formulate questions about logic stories. When students get excited by open-ended questioning, they often come up with innovative questions on their own, thus expanding their capacity for creative thinking and problem solving.

Learning to form, pace, and interpret good questions are vital skills in classrooms moving toward the Standards. Good questions and answers support inquiry in a classroom; poor questions create an atmosphere that stifles risk-taking. Good questions provide ongoing assessment, helping our decisions about instruction be more effective. And as a questioning teacher, we become a valuable model for our students.

Resources for the Road

Blosser, Patricia E. (1991). How To Ask the Right Questions. Arlington, VA: National Science Teachers Association (NSTA).

Burgreen, Sid. (1995, September). How To Annoy Students and Influence Contest Judges. *Science and Children, 33* (1), 28–30.

Chahrour, Janet. (1994, October). Perfecting the Question. *Science Scope, 18* (2), 9–11.

DeCoster, Patricia. (1995, February). Questioning Answers. *The Science Teacher, 62* (2), 46–49.

Gilbert, Steven W. (1992, December). Systematic Questioning. *The Science Teacher, 59* (9), 41–46.

Hinton, Nadine K. (1994, February). The Pyramid Approach to Reading, Writing, and Asking Questions. *Science Scope, 17* (5), 44–46.

Keys, Carolyn W. (1996, February). Inquiring Minds Want To Know. *Science Scope, 19* (5), 17–19.

Kulas, Linda Lingenfelter. (1995, January). I Wonder... *Science and Children, 32* (4), 16–18, 32.

Otto, Paul B. (1991, April). Finding an Answer in Questioning Strategies. *Science and Children, 28* (7), 44–47.

Pearlman, Susan, and Pericak-Spector, Kathleen. (1992, October). Expect the Unexpected Question. *Science and Children, 30* (2), 36–37.

Penick, John E., Crow, Linda W., and Bonnstetter, Ronald J. (1996, January). Questions Are the Answer. *The Science Teacher, 63* (1), 27–29.

Rowe, Mary Budd. (1987). Using Wait Time To Stimulate Inquiry. In W. W. Wilen (Ed.), *Questions, Questioning Techniques, and Effective Teaching* (pp. 95–106). Washington, DC: National Education Association (NEA).

Vincent, Fred V. (1993, May). "What If...? Questions That Stimulate Classroom Discussion. *The Science Teacher, 60* (5), 30–32.

About Teaching Standard D

Designing and Managing the Physical Environment

Moving toward the Standards will require new paradigms in virtually every area of education. With the help of their administrators and community, teachers must explore new ground and redefine the terms "classroom," "school," "class," and "school day."

New Dimensions in Time and Space

To reach the vision of the Standards, we will have to explore new limits of time and space. (For discussions about space, see the Program Standards chapter, page 139.) A significant body of research confirms that time is a key factor in learning. Time on task, time for exploration and reflection, and flexible time for integration are all part of the new science education.

Time on Task. While we know that "perfect attendance" can be unrelated to performance, conversely we also know too well that lost minutes and interruptions negatively impact what happens in our classrooms (see Wang, 1993). We need to be creative in maximizing every minute in a class. This can include everything from packaging and dispensing materials more effectively to using computers to take attendance, post grades, and monitor student performance and resources. Saving one minute a day in each class period can actually add three days of instruction to the year!

Michael Clough and others (1994) suggest that a first step in increasing what they call "engaged time" with students is to discover where time is lost. Their own study uncovered 2.5 minutes at the beginning of each class taking attendance and 5.5 minutes at the end of each class, for a total of 29 periods a year!

They suggest reclaiming this time by using bell work at the beginning of class (when students are already in their seats) and restricting passes to the last 5 minutes of class. Putting papers in student folders saves time lost distributing them, overplanning lessons eliminates dead time, and showing students how to use time effectively increases their time on task.

Time To Learn. Many faculty have discarded the paradigm of 180 days and one-hour classes. Block schedules

---

TEACHING STANDARD D

Teachers of science design and manage learning environments that provide students with the time, space, and resources needed for learning science. In doing this, teachers

- structure the time available so that students are able to engage in extended investigations
- create a setting for student work that is flexible and supportive of science inquiry
- ensure a safe working environment
- make the available science tools, materials, media, and technological resources accessible to students
- identify and use resources outside the school
- engage students in designing the learning environment

Reprinted with permission from the *National Science Education Standards.* © 1996 National Academy of Sciences. Courtesy of the National Academy Press, Washington, D.C.

are becoming the norm rather than the exception in many schools. Using computer networks to monitor where students are and what they are doing removes an important structural barrier to modifying schedules. Clough cites models of 50-, 44-, and 35-minute classes. (Cutting down time spent going from class to class rescues lost time.)

Block scheduling is becoming increasingly popular in laboratory science classes. Terrilee Day (1995) believes that 90-minute blocks four times a week are ideal for chemistry courses. Janet Gerking (1995) describes 90-minute blocks for science every other day.

"Longer class periods allow students extended time for lab work, hands-on projects, field trips, thorough discussions, varied teaching styles, and in-depth study," says Gerking. And planning for the longer blocks is much the same as planning for a series of shorter classes, although students still function best when they are not asked to concentrate on reading or listening for more minutes than their age. (So 15 minutes is about maximum for 9th graders to pay attention to a lecture or to read silently without having discussion breaks.)

Time To Collaborate. Perhaps the key component of the school time revolution has been the incorporation of more time for co-planning and integrated programs in schools. Blocks, flex schedules, late starts, and community activities for students all contribute to greater flexibility in time for teachers.

Resetting the school clock is only one of many pathways toward the Standards. But the process of rethinking the class period is a valuable first step toward challenging all the paradigms in teaching. Isn't it time for us to try something new?

Resources for the Road

Aldridge, Bill G. (1995, October). Invited Paper. *The Science Teacher, 62* (7), 8.

Canaday, Robert Lynn, and Retig, Michael D. (1995, November). The Power of Innovative Scheduling. *Educational Leadership, 53* (3), 4–10.

Clough, Michael P., Smasal, Randal J., and Clough, Douglas R. (1994, September). Managing Each Minute: Strategies for Reclaiming Lost Instructional Time. *The Science Teacher, 61* (6), 30–34.

Day, Terrilee. (1995, April). New Class on the Block: One School's Successful Change to Block Scheduling. *The Science Teacher, 62* (4), 28–30.

DeCoster, Patricia A. (1992, September). Time Saving Tactics. *The Science Teacher, 59* (6), 34–37.

Gerking, Janet L. (1995, April). Building Block Schedules. *The Science Teacher, 62* (4), 23–27.

Hackmann, Donald G. (1995, November). Ten Guidelines for Implementing Block Scheduling. *Educational Leadership, 53* (3), 24–27.

Wang, Margaret C., Haertel, Geneva D., and Walberg, Herbert J. (1993, December/1994, January). What Helps Students Learn? *Educational Leadership, 51* (4), 74–79.

About Teaching Standard E

Building Learning Communities

We cannot journey toward the Standards alone; both science and learning are social activities. Teachers must model respect for diverse ideas and must nurture collaborative skills. In the learning community, teachers give their students a significant voice in the decisions about learning that are made every day.

Communities of Learners

Most of us are still in school for a simple reason—we're good at it. Institutional schooling leans heavily toward the linguistic learning style (and the field-independent analytic learner). Those who are most successful at it are most likely to stay in education. So each new generation teaches as it learned best, and schools remain heavily slanted toward a few learners.

But every class has logical, spatial, musical, kinesthetic, intrapersonal, and interpersonal learners, too. Each of them sees the natural world through perspectives very different from ours. Furthermore, every class has some degree of cultural diversity, and most classes have students for whom any learning at all is a challenge. There is little in our training, background, or inclination to equip us to juggle so many learning needs at once.

Bob Samples (1994) describes the problem: "Traditional teaching methods are largely...cognitive.... During the 1960s...hands-on approaches were added.... Yet little was or is done to include 'entrepreneurs' or 'poetic' students." Making assignments that dovetail with diverse learning styles is often just what is needed to re-instill motivation in the science-shy student or to improve the learning atmosphere in a rowdy class (translation: a class of students whose predominant learning styles are not verbal or mathematical!).

We also need to see science through the eyes of learners from different cultural backgrounds. This requires not only a repertoire of teaching skills but also carefully developed sensitivity to cultural stereotypes and perceived and real inequities. The System Standards call boldly for equity in education and equal opportunity to learn. (See page 162.) In a multicultural classroom, we must look closely at everything we do—the materials we

TEACHING STANDARD E

Teachers of science develop communities of science learners that reflect the intellectual rigor of scientific inquiry and the attitudes and social values conducive to science learning. In doing this, teachers

- display and demand respect for the diverse ideas, skills, and experiences of all students
- enable students to have a significant voice in decisions about the content and context of their work and require students to take responsibility for the learning of all members of the community
- nurture collaboration among students
- structure and facilitate ongoing formal and informal discussion based on a shared understanding of rules of scientific discourse

continued next page

Realizing the Teaching Standards

> • model and emphasize the skills, attitudes, and values of scientific inquiry
>
> Reprinted with permission from the *National Science Education Standards.* © 1996 National Academy of Sciences. Courtesy of the National Academy Press, Washington, D.C.

choose, the examples and words we use, the methods we select, and the assessments we design.

Consider a few examples:

Language. The words used in many science texts may have significantly different meanings for Native American students. The same might be said for students whose families speak a language other than English at home.

Environment. Children from at-risk families often come to school without having had the early exploratory experiences necessary to develop language skills. Without sufficient language ability, these students may be delayed in being able to communicate higher-order, abstract ideas.

Stereotypes. Not only is the stereotype of the white, European, male scientist a barrier to girls and minority students in science, but the stereotype of the highly mathematical Asian student may also be a real hurdle for Asian students with other learning styles.

Modifications. Teachers and students are often so caught up in "fair play" that the concept of modifying class rules for challenged students seems discrepant. But the Standards remind us clearly that while all students will travel the path toward content mastery, their rate and mode of travel may vary. Modifications for special-needs students might include translating text to oral lessons, partitioning units differently, increasing the extent of hands-on discovery, or using a variety of technologies.

How can we support diverse learners while working to build a real community of learners? First, by modeling acceptance and tolerance ourselves. Second, by using the resources of our profession to consciously choose methods and assessments that support learning styles other than the one we are most familiar with—our own. (The best method is a combination of methods.) Perhaps most importantly, we need to recognize that much of learning is a social activity. Cooperative learning techniques involve every learner, encourage tolerance, and create opportunities for students to appreciate one another's perspectives. (*Note:* For more information on equity issues, see Assessment Standard B, page 49; Program Standard E, page 143; and System Standard E, page 162.)

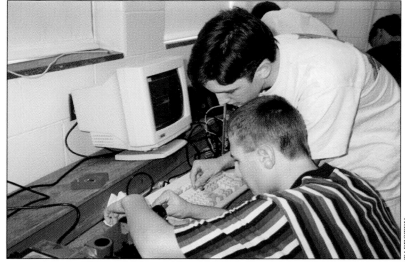

Resources for the Road

Allen-Sommerville, Lenola. (1996, February). Capitalizing on Diversity. *The Science Teacher*, 63 (2), 20–23.

Armstrong, Thomas. (1994). *Multiple Intelligences in the Classroom.* Alexandria, VA: Association for Supervision and Curriculum Development (ASCD).

Atwater, Mary M. (1993, March). Multicultural Science Education. *The Science Teacher*, 60 (2), 33–37.

Atwater, Mary M. (1995, February). The Multicultural Science Classroom. Part I: Meeting the Needs of a Diverse Student Population. *The Science Teacher*, 62 (2), 21–23.

Atwater, Mary M. (1995, April). The Multicultural Science Classroom. Part II: Assisting All Students with Science Acquisition. *The Science Teacher*, 62 (4), 42–45.

Baptiste, H. Prentice, and Key, Shirley Gholston. (1996, February). Cultural Inclusion: Where Does Your Program Stand? *The Science Teacher*, 63 (2), 32–35.

Bernhardt, Elizabeth, Hirsch, Gretchen, Teemant, Annela, and Rodriguez-Munoz, Marisol. (1996, February). Language Diversity and Science. *The Science Teacher*, 63 (2), 25–27.

Bryant, Jr., Napoleon A. (1996, February). Make the Curriculum Multicultural. *The Science Teacher*, 63 (2), 28–31.

Campbell, Melvin, and Burton, VirLynn. (1994, April). Learning in Their Own Style. *Science and Children*, 31 (7), 22–24, 39.

Carey, Shelley Johnson (Ed.). (1993). *Science for All Cultures.* Arlington, VA: National Science Teachers Association (NSTA).

Dorough, Donna K., and Bonner, Robert C. (1996, February). Incorporating Multicultural Dialogue. *The Science Teacher*, 63 (2), 50–52.

Mason, Cheryl L., and Barba, Robertta H. (1992, May). Equal Opportunity Science. *The Science Teacher*, 59 (5), 22–26.

National Education Commission on Time and Learning. (1994). *Prisoners of Time.* Washington, DC: Author.

Ossont, Dave. (1993, May). How I Use Cooperative Learning. *Science Scope*, 16 (8) 28–31.

Peltz, William H. (1990, December). Can Girls + Science - Stereotypes = Success? *The Science Teacher*, 57 (9), 44–49.

Robblee, Karen M. (1991, January). Cooperative Chemistry. *The Science Teacher*, 58 (1), 20–23.

Samples, Bob. (1994, February). Instructional Diversity: Teaching to Your Students' Strengths. *The Science Teacher*, 61 (2), 14–17.

Saunders, Gerry W., Wise, Kevin, and Golden, Tim S. (1995, February). Visual Learning. *The Science Teacher*, 62 (2), 42–45.

Simons, Grace H., and Hepner, Nancy. (1992, September). The Special Student in Science. *Science Scope*, 16 (1), 34–39, 54.

Thorp, Steve (Ed.). (1991). *Race, Equality, and Science Teaching.* Hatfield, Herts. UK: Association for Science Education (ASE).

Watson, Scott B. (1992, February). Cooperative Methods. *Science and Children*, 29 (5), 30–31, 47.

Weld, Jeffrey D. (1990, November). Making Science Accessible. *The Science Teacher*, 57 (8), 34–38.

About Teaching Standard F

School Planning

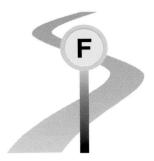

No one but teachers can design and implement school science programs. The design of courses and the allocation of time and resources can best be determined by those who will carry out the plans. Teachers must also plan and implement the professional growth experiences they will need to succeed.

Rewriting the Course Syllabus

To make the journey to the Standards, we will have to shed much of the baggage that has burdened us throughout the 20th century. First and perhaps foremost, the layer-cake course structure for school science that was first proposed by the so-called "Committee of Ten" at the turn of the century, with one year in each of the upper grades devoted to each of the science disciplines, must be packed away with the slide rule and the Canada balsam. New content requires new courses laid out in very different sequences. Consider some of the changes that professionals have already piloted.

New Sequencing. Today's biology has little to do with the parts of protozoa, and tomorrow's earth science integrates advanced physical science and mathematics. The traditional sequence of earth science in junior high, then biology, then (in the best cases) chemistry and physics, is simply unteachable today. New courses have been designed to enable students to explore the most up-to-date scientific thinking. These include

- *Biological Science: A Human Approach:* Introductory coursework in life science focusing on what early adolescents are most interested in—themselves!
- *Activities for the Changing Earth System:* Integrated, advanced treatment of earth science for juniors and seniors who have already had chemistry and physics.
- *Active Physics:* Alternative physics course for students with limited math and reading skills. (see page 84).

New Relevance. Once we have reset our sights on teaching toward adult life and employability, much of the coursework we have cherished for so long becomes irrelevant. Courses permeated with real-world applications have proven not only exciting but also capable of fostering high student achievement.

TEACHING STANDARD F

Teachers of science actively participate in the ongoing planning and development of the school science program. In doing this, teachers
- plan and develop the school science program
- participate in decisions concerning the allocation of time and other resources to the science program
- participate fully in planning and implementing professional growth and development strategies for themselves and their colleagues

Reprinted with permission from the *National Science Education Standards.* © 1996 National Academy of Sciences. Courtesy of the National Academy Press, Washington, D.C.

These include
- Principles of Technology: A program that covers fewer traditional physics concepts and makes extensive use of labs

borrowed from industry.
- Applied Biology/Chemistry: A program that integrates environmental health and biotechnology applications from industry.
- Chemistry in the Community: Offers high relevance and higher-order thinking for students through a liberal dose of real-world applications.

New Organization. Teachers around the country are piloting courses that bear very little resemblance to those that have been around for 90 years. Integrated science takes many forms, but almost every pilot program has yielded high student performance. The barriers between the disciplines, it seems, are more in the minds of instructors than students—and certainly are nonexistent in the natural world.
- NSTA's Scope, Sequence and Coordination of Science: This curriculum approach includes every science every year, coordinated and sequenced from the concrete to the abstract in tandem with student development. The program has made a strong argument for integrated science in pilot programs across the nation. Evaluations have been positive, and institutional support strong. Strongest applications to date have been in grades 9 and 10 before students begin to make choices about careers and specialization.

In schools that haven't moved to integrated science, the chief barrier seems to be teacher training and certification. Veterans who were credentialed several decades ago may not be able to teach integrated courses. Despite this, innovative systems rely on team- or co-teaching, which allows teachers to share expertise.

Lean and Mean? Probably the most dramatic shift that we face in school coursework is the need to reduce the amount of content in science courses. "Less is more" isn't just a buzzword, but a real milepost for making virtually every course decision on our pathway toward the Standards.

Resources for the Road

Aldridge, Bill G. (Ed.) (1996). *Scope, Sequence, and Coordination: A High School Framework for Science Education.* Arlington, VA: National Science Teachers Association (NSTA).

American Association of Physics Teachers (AAPT). Active Physics. (1996). College Park, MD: Author.

American Chemical Society (ACS). ChemCom: Chemistry in the Community. (2nd ed.). Dubuque, IA: Kendall Hunt.

Biological Sciences Curriculum Study (BSCS). (1995). *BSCS Biology: A Human Approach.* Dubuque, IA: Kendall Hunt.

Center for Occupational Research and Development. Principles of Technology Series Modules. (1990). Waco, TX: CORD Communications.

Center for Occupational Research and Development. Applied Biology/Chemistry. (1992). Waco, TX: CORD Communications.

Fortner, R. W., and Mayer, Victor J. (1993). *Activities for the Changing Earth System.* Columbus, OH: The Ohio State University.

Changing Emphases

The National Science Education Standards envision change throughout the system. The Teaching Standards encompass the following changes in emphases:

LESS EMPHASIS ON	MORE EMPHASIS ON
Treating all students alike and responding to the group as a whole	Understanding and responding to individual student's interests, strengths, experiences, and needs
Rigidly following curriculum	Selecting and adapting curriculum
Focusing on student acquisition of information	Focusing on student understanding and use of scientific knowledge, ideas, and inquiry processes
Presenting scientific knowledge through lecture, text, and demonstration	Guiding students in active and extended scientific inquiry
Asking for recitation of acquired knowledge	Providing opportunities for scientific discussion and debate among students
Testing students for factual information at the end of the unit or chapter	Continuously assessing student understanding
Maintaining responsibility and authority	Sharing responsibility for learning with students
Supporting competition	Supporting a classroom community with cooperation, shared responsibility, and respect
Working alone	Working with other teachers to enhance the science program

Reprinted with permission from the *National Science Education Standards*. © 1996 National Academy of Sciences. Courtesy of the National Academy Press, Washington, D.C.

Professional Development Standards

The Standards challenge us to become dedicated, lifelong learners.

Reaching for the Professional Development Standards

Every time we see learning reflected in our students' eyes, we become better teachers. Like children, we construct new models by seeing the results of our efforts. If the goals of our teaching are clear and we monitor the assessments we give, we will continue to improve through the same inquiry methods we foster in our students. This is the most basic professional development program for every teacher.

But the pathways for professional development go far beyond practical experience. The Standards urge us to improve our knowledge of content in the traditional science disciplines; our understanding of learning theory; and the specialized knowledge that integrates content, curriculum, learning, teaching, and students (called pedagogical content knowledge). Tomorrow's professional development must reflect sound principles of research in all these fields.

A Personal Pathway

The Standards challenge us to become dedicated, lifelong learners, to use "district goals and expectations...to set personal goals for professional development...." And the Standards place the responsibility for designing a personal pathway squarely on the shoulders of the teachers, *not* administrators, school boards, or systems.

This will be new territory for many of us, who in the past may have waited for our district or local university to "offer a program." It will also be new ground for school administrators who have often assumed that a single mode of staff inservice was right for every teacher. But the world is changing—and so are the lifelong learning needs of individuals.

The Professional Development Standards include four distinct areas: science content and pedagogical content knowledge (what we need to know) and teacher attitudes and program design (how we get there). Despite this simple outline, the pathway to professional development is a marathon, not a 100-yard dash. The winners among us will be those who realize that the finish line is far in the distance and who train and persevere to reach their goal. They will be the ones who dedicate themselves to finding out more about teaching and learning over an entire career.

The What—Content We Need To Know

The first two standards in the area of Professional Development describe essential content and elements of teaching. Adequate content is not defined by credit hours, but described as a "broad base of scientific understanding" in the science disciplines and in learning theory.

For K–12 generalists, in-depth experiences in at least one science discipline will be needed to provide a sense of the process of inquiry. Middle level teachers, the Standards suggest, must have sufficient understanding "to elicit student understandings and beliefs about scientific ideas and to use these data to formulate activities that will aid in the development of sound scientific ideas"—a significant challenge. Secondary teachers are urged to have a background in *all* science disciplines and sufficient depth in their area of focus "to seek and find new information and to inquire skillfully within that science."

The Standards validate a movement that began with the accreditation standards for college programs developed by the National Council for the Accreditation of Teacher Education (NCATE) and with the NSTA Teacher Certification Program, which defined a basic content core so broad that teacher preparation often takes five years.

The Standards suggest that every teacher of science should carry out real scientific research, experience societal issues firsthand, and know how to access information and how to continue learning science. The breadth of knowledge implied in most content programs also includes competence in computer technology and mathematics through calculus.

The Standards also give credence to the vast, growing body of research on learning theory and pedagogy. Rather than just the ability to convey knowledge, teaching is now seen as an extremely complex task. Knowledge of the learner—cognitive, social, and psychological—is key and varies depending upon the age of the student. Such knowledge is absolutely critical to making decisions about what will be taught, how instruction will be paced, and which activities will engage students.

Likewise, we need to know and be able to use basic strategies, such as hands-on activities, questioning, discussion, and cooperative learning, and have the ability to assess prior knowledge. Finally, knowledge of curriculum and assessment materials is necessary, along with the ability to adapt materials to the needs of a particular student.

Such a broad, dual-focus preparation implies a significant change. To attract the best into such rigorous programs, the career path in teaching may have to change significantly. For those reasons, the Professional Development Standards have great implications for the parallel guidelines spelled out in the System Standards.

The How—Pathways to Excellence

Knowledge of science and science teaching are necessary but not sufficient. Another component in professional development is the ability to integrate this knowledge into an attitude of lifelong learning that we can model for our students. This skill demands a thinking, caring, well-educated, professional science teacher who never stops learning. We might be guided by the following:

- Professional development is a career-long endeavor. A degree does not make one an excellent practitioner. Constant learning must be a major career theme, with content knowledge and pedagogy enriched on a regular basis.
- We are responsible for our *own* professional development. We must take ownership for planning our career-long program and periodically evaluate our progress. Professional development should be part of district evaluations and staff development plans.
- We are not just targets of professional development, but can and should play the role of source and supporter at various times in our careers. We have much to contribute both as leaders and mentors.
- One size does *not* fit all. The path to excellence may not be the same for each of us. No one model will fit even a majority of teachers. Local support must be flexible.

There are so many ways to improve teaching skills that any list will be inadequate. But despite this caution, let's take a closer look at a few options:

Graduate Courses. One of the most common routes to improve skills, graduate courses offer structured opportunities to practice laboratory, research, and teaching skills. Some programs give us the flexibility to design our own experiences toward a degree. Many school systems offer a salary increase when we receive an advanced degree, providing a strong bias toward this pathway. However, programs that are expensive or require long hours away from the daily work of teaching are less desirable.

Professional Associations. Tens of thousands of us are "addicted" to our professional associations and the opportunities they provide for professional growth. Many of us were first attracted by a conference or two. For the cost of registration, we found we could select from hundreds of workshops and sample expansive exhibits. Soon we found ourselves networking with other teachers and learning from them. Then someone asked us to present a session. The professional "high" from this experience built confidence and often led to other contributions.

Professional Journals. With membership in a professional association, of course, comes the bonus of professional journals, newspapers, and newsletters, which bring classroom ideas and news to our doorstep between conferences. We try some activities we read about and they work well; others we decide to modify. Sometimes a modification is original, and we may even begin submitting articles ourselves. Leadership responsibilities may follow. All of these are growth experiences.

Structured Inservice Programs. Today most districts allow staff committees to design department, building, or district inservice options, with the district providing financial support. If care is taken to avoid requiring all teachers to attend the same program, in-house programs can be very valuable. The chief advantage is that teachers from the same school share the same experiences and can exchange ideas as they practice new techniques.

The most important caution in designing a program for your school is to avoid one-shot programs because they seldom have any effect on instruction. The best inservice programs set a long-range target (a measurable student outcome) and design a series of steps toward that target that include staff development, classroom implementation, focus-group evaluation, and student evaluation. Inservice goals should generally extend from one to three school years in order to validly document measurable effects.

Collaboration with Other Professionals. Teaching is essentially a very private endeavor, but collaboration is inevitably enriching. Observation by peers is an uncomfortable experience for most of us who find the classroom door an effective insulator against possible criticism. But nonthreatening coaching by a fellow science teacher can offer the opportunity to discover other styles, strategies, and options in teaching science. Mentoring (whether formal or informal) is another form of collaboration that opens the door for giving and receiving feedback about what and how well we are doing.

Self-Reflection and Inquiry. Not all learning experiences need to involve structured meetings, nor do they need to be group-oriented. There is much value in reading, studying, and exploring on our own. In fact, an important component of any career-long professional development plan must be self-reflective inquiry activities. As we try out new ideas, we might use journals, audiotapes, and videotapes to track our progress. Many graduate programs encourage action research in classrooms and are willing to supply statistical and pedagogical support.

Evaluation. Using evaluation as a vehicle for professional development will seem strange to most of us for whom the experience of a nonscience teacher in a judging role may seem intimidating. But new, more reflective systems for evaluation can become valuable opportunities for us to communicate the what, the why, and the results of our educational decisionmaking to another educator. If evaluation is to be a tool for growth, however, we must be full partners in developing the evaluation system.

School Improvement for Professional Growth. Setting common goals can provide a school- or district-wide impetus for professional development. A shared direction (such as providing more varied course offerings to all students) can prompt a group to determine what strategy or plan has the best hope of success. Most plans include regular evaluation of progress. Having a voice in the policies and procedures that affect our classrooms will certainly increase our sense of ownership and dedication.

Other Possibilities. Sabbatical leave can give us time to offer professional development to our colleagues. Internships in industry or research provide new perspectives for science teachers. Travel (either to scientific sites or to other classrooms) has long been considered one of the most motivating professional development resources.

Closer to home, reading is perhaps the easiest way to expand knowledge. Exploration of the Internet or other online networks can yield a wealth of resources as well as the opportunity to share ideas with colleagues.

The possibilities are limited only by our own creativity. The critical question is "How does this experience move me toward my personal goal of becoming an *excellent* teacher?"

A Caution and Call for Realism

The list above does not try to catalog every possible method of professional development, but rather describes some general types as an impetus to self-reflection and exploration. But a caution is in order: The path to fulfilling the expectations of the Professional Development Standards is neither direct nor easy. As the Standards themselves state, "Teachers of science will stumble, wrestle, and ponder, while realizing that failure is a natural part of developing new skills and understanding." Human mistakes, blind alleys, changing perspectives, and unexpected inspirations will continue to alter the course of our professional development. We must believe that no endeavor is wasted and no undertaking is devoid of learning.

As we journey toward the vision of the four Professional Development Standards in the following pages, we will explore these pathways:
- learning science content
- learning to teach science
- learning to learn
- program development

About Professional Development Standard A

Learning Science Content

Teachers of science must understand a large body of science content, learned through the perspectives and methods of inquiry.

Scientist in the Classroom

The volume of knowledge in the world doubles every six months. Yet the average biology teacher graduated from college 14 years ago. When she or he received the degree, genes were mapped through crossover analyses, vaccines were made in live animals, and AIDS was an obscure curiosity in the Centers for Disease Control labs.

Perhaps more important, when most of today's biology teachers received their baccalaureate degrees, the subject could be understood without a having minor in chemistry *and* physics! Yesterday's chemistry and physics teachers *thought* they were studying distinct disciplines. And Earth science teachers concentrated on the study of geology.

The Standards demand that science teachers understand content—not only as a subject to be mastered but also from the perspective of a scientist. They suggest that every science teacher should have conducted research personally and be able to do so again confidently (even if the instrumentation or the techniques race ahead of experience!).

To become a secondary science teacher, the Standards suggest that preparation should include significant training in *every* science discipline with an underlying foundation of mathematics and competence in computer technology. This is certainly a daunting agenda.

The content standards—and the preparation they require—raise two significant challenges for school systems across the nation. First, if the coursework required prior to beginning teaching is roughly equivalent to that required of those majoring in science, how will colleges attract the best graduates? Will salaries, schedules, and, most important, the social status of teachers, be commensurate with the preparation being required?

PROFESSIONAL DEVELOPMENT STANDARD A

Professional development for teachers of science requires learning essential science content through the perspectives and methods of inquiry. Science learning experiences for teachers must

- involve teachers in actively investigating phenomena that can be studied scientifically, interpreting results, and making sense of findings consistent with currently accepted scientific understanding
- address issues, events, problems, or topics significant in science and of interest to participants
- introduce teachers to scientific literature, media, and technological resources that expand their science knowledge and their ability to access further knowledge

continued on next page

- build on the teacher's current science understanding, ability, and attitudes
- incorporate ongoing reflection on the process and outcomes of understanding science through inquiry
- encourage and support teachers in efforts to collaborate

Reprinted with permission from the *National Science Education Standards.* © 1996 National Academy of Sciences. Courtesy of the National Academy Press, Washington, D.C.

The second challenge is for continued professional development in the content area. Just reading about advances in biology or Earth science can't keep us up to date. Summer experiences in *real* laboratories using new instrumentation and new techniques will be essential. (You can find lists of summer opportunities in *NSTA Reports!.)* But how can systems encourage this sort of grueling "vacation" for science teachers? Conversely, if teachers spend summers in research labs or in industry, will schools have difficulty luring them back?

What about our own *individual* paths to becoming better prepared in our content area? We mustn't let the high level of preparation suggested discourage us from taking a small step in the direction of more modern understandings. For us as inservice teachers, opportunities for content update abound:

- Try a content infusion. Read *Science News, Discover, Scientific American,* the weekly "Science Times" section of the *New York Times,* or other publications of interest. Attend the Shell Science Seminars at NSTA National Conventions. Or you might even consider going to a scientific convention, such as meetings of the American Association for the Advancement of Science or the American Chemical Society.
- Access online information sources, and download updates in areas that you are teaching.
- Apply for a summer inservice in a college laboratory. (Forget the seat time; try a lab bench.)
- Try an internship in an industrial research laboratory.
- Volunteer for a field research project in your discipline.

Take the plunge. Feel like a scientist again—the spirit will infect your whole class next fall.

Resources for the Road

Anderson, Norman D. (1993). *Scientists: Tips for Making Presentations to Teachers.* Raleigh, NC: North Carolina State University, SCI-LINK/GLOBE-NET.

Fraser-Abder, Pamela, and Leonhardt, Nina. (1996, January). Research Experiences for Teachers. *The Science Teacher, 63* (1), 30–33.

Massell, Laura Nault, and Searles, Georgiana M. (1995, February). An Alliance for Science. *Science and Children, 32* (5), 22–25.

Moffit, Mary, Friesema, Jane, and Brady, Mary. (1994, December). Bringing Teachers Up to Speed. *The Executive Educator, 16* (12), 16–17, 47.

National Science Teachers Association (NSTA). (1992). NSTA Standards for Science Teacher Preparation. Arlington, VA: Author. Adopted by the National Council for the Accreditation of Teacher Education (NCATE).

About Professional Development Standard B

Learning To Teach Science

Knowledge of science must be integrated with knowledge of learning, pedagogy, and students and with the development of the skills to apply this knowledge in the classroom.

The Science of Students

Rodney Dangerfield could certainly do a routine about cognitive research: For generations, the science of learning has had a difficult time getting respect. Yet the body of research about how students learn has gradually become the foundation of teaching. Certainly, teaching is still an art—but today, it is also one of the "hard" sciences. Researchers are close to discovery their own unified field theory of learning in the body of results called constructivism.

The way that experienced teachers understand learning sometimes seems intuitive; they seem to anticipate misunderstandings and plan educational experiences at critical points to minimize them. But teachers with very different methods often have had similar results in terms of student achievement. How do you teach another professional how to do what you do? Learning theory suggests that the science is not in "what?" but in "why?"

The specialized knowledge of the science teacher (pedagogical content knowledge) is distinctly our own. We need to know
- how to prepare, package, schedule, manage, and assess laboratories
- how to choose experiences that are developmentally appropriate
- how to blend direct and indirect instruction appropriately

To encourage a greater depth of understanding in cognition and pedagogical content knowledge, preservice teachers are now doing more field practicums coupled to classes that analyze the "what" and "why" of their observations.

But how about those of us inservice teachers who have had reasonably good success teaching for a decade or more? How can good teachers become better? Often the answer is in *action research*. To move from good to great, we can apply the principles of the

PROFESSIONAL DEVELOPMENT STANDARD B

Professional development for teachers of science requires integrating knowledge of science, learning, pedagogy, and students; it also requires applying that knowledge to science teaching. Learning experiences for teachers of science must
- connect and integrate all pertinent aspects of science and science education
- occur in a variety of places where effective science teaching can be illustrated and modeled, permitting teachers to struggle with real situations and expand their knowledge and skills in appropriate contexts
- address teachers' needs as learners and build on their current knowledge of science content, teaching, and learning

continued on next page

> • use inquiry, reflection, interpretation of research, modeling, and guided practice to build understanding and skill in science teaching
>
> Reprinted with permission from the *National Science Education Standards*. © 1996 National Academy of Sciences. Courtesy of the National Academy Press, Washington, D.C.

science of learning to our own classrooms.

Action research begins when we focus on a single outcome and a single standard for evaluating the effectiveness of instruction toward that outcome. The next step is often a partnership with a university or resource center for technical assistance in identifying assessment instruments and analyzing data. Using the principles of scientific inquiry to isolate variables, we look for the most effective methods in our classroom. Sometimes the results become part of a larger study or a meta-analysis of similar studies; other times action research is simply part of our own professional development repertoire.

Action research brings today's cognitive science "up close and personal" in our classrooms. Significant results boost our effectiveness and expand our profession's body of knowledge. But even when results are inconclusive, the practice of action research makes cognitive research relevant.

Districts can encourage action research in classrooms by providing technical assistance in data analysis of such questions as these: Could standardized tests in the next grade be analyzed to show the achievement of a single teacher in a single objective area *(without* threatening the investigator)? Could new assessment methods be made available, scored, or processed electronically? Districts might also support us in communicating our research efforts to parents in a nonthreatening way.

With the acceptance of cognitive research as a rigorous field of science, our need to see the "science" in what we do grows stronger every day. Just like our students, we learn best by doing—by experimenting and applying our new knowledge in our own classrooms. Action research is just one pathway toward the familiarity with learning theory and pedagogical content knowledge defined by the Standards.

ROBERT D. DAVIS

Resources for the Road

Clough, Michael. (1992, October). Research *Is* Required Reading: Keeping Up with Your Profession. *The Science Teacher, 59* (7), 36–39.

Kennedy, Mary M. (1990). Teachers' Subject Matter Knowledge (ERIC Document Reproduction Service No. ED322100).

About Professional Development Standard C

Learning To Learn

Professional development provides the knowledge, skills, and attitudes for lifelong learning. It includes collegial self-examination, opportunities for feedback, opportunities for sharing expertise and accessing resources, and research.

With a Little Help from Our Friends

Where is the best source of professional development for you? It's probably just around the corner in a classroom near you. Often a fellow teacher offers the best inspiration and the best mirror of performance.

Schools that want to increase collaboration often schedule co-planning time without structuring it at first. Scheduling middle level students in teams and providing all their teachers with the same preparation period is an open invitation to teaming. So is coordinating the preparation period of same-grade teachers. Districts have often enhanced awareness of problems by simply arranging release time for teachers to observe another grade. Such days go a long way toward reducing the "why didn't they teach X?" syndrome across grade levels.

The next step along the path to collaboration is to identify themes or assessment goals and ask groups of teachers to lead strategic planning efforts toward those goals. A simple twist is the "share-a-thon" meeting. It can be as simple as a handful of teachers bringing their favorite activity to the lunchroom, or as elaborate as a district-wide poster session. The *National Science Education Standards* might be a useful tool in such sessions.

When planning collaboration, care should be taken to arrange meetings at a time when enthusiasm will be high. After-school sessions are usually less successful than early-morning sessions. If after school is the only option, a snack or meal helps increase camaraderie. Another caution: Avoid having one partner or group take over the collaboration effort. Elementary teachers may have as much to teach about pedagogy as secondary teachers may about science content.

Peer coaching can be another path to professional de-

PROFESSIONAL DEVELOPMENT STANDARD C

Professional development for teachers of science requires building understanding and ability for lifelong learning. Professional development activities must

- provide regular, frequent opportunities for individual and collegial examination and reflection on classroom and institutional practice
- provide opportunities for teachers to receive feedback about their teaching and to understand, analyze, and apply that feedback to improve their practice
- provide opportunities for teachers to learn and use various tools and techniques for self-reflection and collegial reflection, such as peer coaching, portfolios, and journals

continued on next page

- support the sharing of teacher expertise by preparing and using mentors, teacher advisors, coaches, lead teachers, and resource teachers to provide professional development opportunities
- provide opportunities to know and have access to existing research and experiential knowledge
- provide opportunities to learn and use the skills of research to generate new knowledge about science and the teaching and learning of science

Reprinted with permission from the *National Science Education Standards*. © 1996 National Academy of Sciences. Courtesy of the National Academy Press, Washington, D.C.

velopment. Here teachers help other teachers using techniques based on models of clinical supervision. This is *never* an evaluation; it begins with an invitation, and its conclusions are nonjudgmental.

In science, peer coaching often involves having one teacher gather data for another on a specific parameter that affects performance. It's not just a visit or an observation. The essential element in a peer-coaching experience is asking a specific question that can be answered through the coaching experience. For example, a teacher may be concerned about eye contact, questioning, wait time, body movement, clarity of explanations, or student involvement.

Once a question is identified, the teachers usually work together to perfect a data-recording method. "Scripting" class interactions takes practice; students move quickly, and care must be taken to avoid subjective interpretations. But once a method is agreed to and perfected, the presence of a peer coach can be invaluable. Another teacher is the best second set of eyes we can have.

Mentoring is a variation on peer coaching that often pairs master or veteran teachers with those newer to the field. Mentor teachers act as role models, enriching the methods repertoire of the novice and providing a sounding board for the inevitable frustrations of being new to the profession. In the best of situations a mentor is a walking survival kit; she or he answers all the questions that a new teacher has but is afraid to go to the office to ask.

College–school partnerships are a third twist on peer partnerships. Pairing with a college teacher works best when both partners see advantages in the collaboration: The college instructor is rejuvenated through experiences with K–12 students, and the classroom teacher gains access to the resources and technology that most college staff take for granted. College–school partnerships often begin informally; college-instructor parents offer to help in their children's schools, or inservice teachers in graduate programs stay in contact with instructors after a course has ended. Several grant-funded projects provide seed money for such partnerships, but a lack of funding shouldn't stop anyone from piloting a cooperative partnership.

A fourth form of peer coaching takes an electronic path. Today it's easy to connect with one or many teachers in the same field on the Internet (including through the NSTA homepage at http://www.nsta.org) or through an online service. Such links are invaluable when we run into a class-specific question: How long can a Biuret solution be shelved? Why did the color change in the oxidation reaction take so long? How can I dispose of hematoxylin? We might ask a school policy question: Does anyone have advice on collecting lab fees?

Barriers to Collaboration

If collaboration is so valuable, then why don't we all seek it out? Inger (1995) suggests that there are three main barriers: norms of privacy (Teaching is "entrepreneurial," and teacher autonomy is grounded in the principle of noninterference.); subject affiliation (Teachers at the high school level view themselves as subject matter specialists, and the insularity of classrooms strengthens the stereotype.); and barriers between vocational and academic teachers (Separated by funding and purpose, these two disciplines form separate worlds in comprehensive high schools.).

But despite the barriers, Inger suggests that schools and systems can make collaboration work by endorsing and rewarding partnerships; organizing schools into teams; giving teachers latitude to influence curriculum and instruction; and giving time, training, and material support.

Today, teaching is complex and complicated by systemic factors that cannot be defined in methods textbooks. Other teachers offer us the best sources of support on the pathway to the Standards.

Resources for the Road

Bieda, Robert, Gibbs, Richard, and Goldie, Susan. (1990, January). Collaborate with Your Collegiate Colleagues. *The Science Teacher,* 57 (1), 40–42.

Bowman, Ray D. (1991, September). Encouraging Excellence: Forging a Link with Your Collegiate Colleagues. *The Science Teacher,* 58 (6), 62–64.

Glickman, Carl D. (Ed.). (1992). *Supervision in Transition: 1992 Yearbook of the Association for Supervision and Curriculum Development.* Alexandria, VA: ASCD.

Inger, Morton. (1993). Teacher Collaboration in Urban Secondary Schools. ERIC/CUE Digest No. 93 (ERIC Reproduction Service Document No. ED363676).

Joyce, Bruce (Ed.). (1990). *Changing School Culture Through Staff Development: 1990 Yearbook of the Association for Supervision and Curriculum Development.* Alexandria, VA: ASCD.

Koballa, Jr., Thomas R., Eidson, Sandra D., Finco-Kent, Deborah, Grimes, Stanley, Kight, Carol R., and Sambs, Hermann. (1992, September). Peer Coaching. *The Science Teacher,* 59 (6), 42–45.

Kroto, Joseph J. (1993, September). Reach Out to Your Rookies. *The Science Teacher,* 60 (6), 49–52.

Rodrigue, Polly, and Tingle, Joy B. (1994, January). The Extra Step: Linking Inservice and Preservice Teachers. *Science and Children,* 31 (4), 34–36.

Rudolph, Sidney, and Preston, Linda. (1995, September). Teaching Teachers. *The Science Teacher,* 62 (6), 30–32.

Voyles, Martha, and Charnetski, Deborah. (1994, March). A Powerful Partnership. *Science and Children,* 31 (6), 25–27, 46.

Wallace, John, and Louden, William. (1995, February). Making a Case for Peer Review. *The Science Teacher,* 62 (2), 39–41.

About Professional Development Standard D

Designing Professional Development Programs

Preservice and inservice programs should be coherent and integrated; have clear goals and individualized options; and be attuned to the history, culture, and organization of each school. Continuous assessment supports program improvement.

Planning for Lifelong Learning

Three powerful ideas are guiding today's staff development, according to Dennis Sparks (1995): results-driven education, systems thinking, and constructivism. As a result, staff development in the 1990s looks different from the programs we've known in the past. Today's programs

- emphasize *both* individual and organizational development
- are driven by a clear, coherent strategic plan (not fragmented as in the past)
- are focused by schools, not districts
- concentrate on student needs (not adult needs)
- are job-embedded, not centered at remote sites
- let teachers study themselves rather than turning the task over to "visiting experts"
- combine generic and content-specific skills
- have staff developers focus on facilitation rather than training
- share responsibility for staff development among

There's no cookbook for professional development because the recipe for every school and every teacher is different. The history and needs of each staff, the resources of each community, and the financial base of each school will all influence the choices that are made.

While no one can define the exact route we and our peers will take, there are many road signs along the way.

PROFESSIONAL DEVELOPMENT STANDARD D

Professional development programs for teachers of science must be coherent and integrated. Quality preservice and inservice programs are characterized by

- clear, shared goals based on a vision of science learning, teaching, and teacher development congruent with the *National Science Education Standards*
- integration and coordination of the program components so that understanding and ability can be built over time, reinforced continuously, and practiced in a variety of situations
- options that recognize the developmental nature

continued on next page

Reaching for the Professional Development Standards 37

of teacher professional growth and individual and group interests, as well as the needs of teachers who have varying degrees of experience, professional expertise, and proficiency
- collaboration among the people involved in programs, including teachers, teacher educators, teacher unions, scientists, administrators, policymakers, members of professional and scientific organizations, parents, and businesspeople, with clear respect for the perspectives and expertise of each
- recognition of the history, culture, and organization of the school environment
- continuous program assessment that captures the perspectives of all those involved, uses a variety of strategies, focuses on the process and effects of the program, and feeds directly into program improvement and evaluation

Reprinted with permission from the *National Science Education Standards*. © 1996 National Academy of Sciences. Courtesy of the National Academy Press, Washington, D.C.

administrators and other leaders
- urge all staff, not only teachers, to focus on continuous improvement
- consider staff development indispensable—certainly no longer a frill

If your school doesn't already have quality professional development, the first steps may be the hardest. Dennis Sparks (1995) suggests that we begin by assessing our knowledge and skills, joining a professional association, working with others to identify relevant activities, becoming active in local reform efforts, and considering seeking certification from the National Board for Professional Teaching Standards or NSTA.

The Professional Development Standards draw for us a general pathway for lifelong growth. But each of us will chart our own way among the many possible options. Bon voyage!

Resources for the Road

ERIC, U.S. Department of Education. Professional Development. (Theme Issue) (1995, Winter). *ERIC Review, 3* (3), 1–32.

Glass, Lynn, Aiuto, Russell, and Andersen, Hans O. (1993). *Revitalizing Teacher Preparation in Science: An Agenda for Action.* Arlington, VA: National Science Teachers Association (NSTA).

Holmes Group. (1990). *Tomorrow's Schools: Principles for the Design of Professional Development Schools.* East Lansing, MI: Author.

Houghton, Mary, and Green, Paul. (1995). *Professional Development for Educators: New State Priorities and Models.* Washington, DC: National Governors' Association.

Mason, Cheryl. (1993). *Preparing and Directing a Teacher Institute.* Arlington, VA: National Science Teachers Association (NSTA).

Sparks, Dennis. (1995, Winter). A Paradigm Shift in Staff Development. In Professional Development. (Theme Issue) *ERIC Review, 3* (3), 2–4.

U.S. Department of Education. (1993). *Improving the Teaching of Science: Staff Development Approaches.* Washington, DC: Author.

Wood, Fred, et al. (1993). How To Organize a School-Based Staff Development Program (ERIC Document Reproduction Service No. ED360752).

Changing Emphases

The National Science Education Standards envision change throughout the system. The Professional Development Standards encompass the following changes in emphases:

LESS EMPHASIS ON	MORE EMPHASIS ON
Transmission of teaching knowledge and skills by lectures	Inquiry into teaching and learning
Learning science by lecture and reading	Learning science through investigation and inquiry
Separation of science and teaching knowledge	Integration of science and teaching knowledge
Separation of theory and practice	Integration of theory and practice in school settings
Individual learning	Collegial and collaborative learning
Fragmented, one-shot sessions	Long-term coherent plans
Courses and workshops	A variety of professional development activities
Reliance on external expertise	Mix of internal and external expertise
Staff developers as educators	Staff developers as facilitators, consultants, and planners
Teacher as technician	Teacher as intellectual, reflective practitioner
Teacher as consumer of knowledge about teaching	Teacher as producer of knowledge about teaching
Teacher as follower	Teacher as leader
Teacher as an individual based in a classroom	Teacher as member of a collegial professional community
Teacher as target of change	Teacher as source and facilitator of change

Reprinted with permission from the *National Science Education Standards*. © 1996 National Academy of Sciences. Courtesy of the National Academy Press, Washington, D.C.

Assessment Standards

***Assessment isn't the final step
in education—it's the first.***

Exploring the Assessment Standards

Assessment isn't the final step in education—it's the first. Without continual reality checks, constructing new knowledge is impossible. There is no point in wondering whether or not we should teach to the test because teaching and assessment are inseparable partners. Like "cart" and "horse," one does not come before the other. As the Standards clearly state, testing *is* learning.

The strong pressure at the national level for new types of assessment can be intimidating for the vast majority of us who survived a single dose of "Evaluation 101" in college. We find it far easier to associate tests with student grades, the most visible measure of what is happening in our classrooms, than with the theory and statistics that we learned in college. Given our limited backgrounds in assessment and intense community pressure for objectivity, it's not surprising that we are nervous about changing how we assess learning.

Yet the vast majority of us already have the tools we need to improve classroom assessment. First of all, we understand our students. We unconsciously make evaluative decisions all the time as we move around the classroom: "She knows it; he needs more practice." An instructor who might pale when asked to conduct a clinical interview or construct an authentic assessment in fact successfully assesses his or her students dozens of times each day.

Second, as teachers of science, we understand the most important principle of the experimental method: "Check one variable at a time." This rule not only guides the simplest classroom experiments but also the most sophisticated statistical techniques in evaluation.

When test makers use analysis of variance to find out whether a test is valid (that is, is it actually testing what it claims to be testing?), they are trying to determine whether a single variable—the learning outcome—is the source of most of the variation found in student test scores. Mathematical analyses can sometimes determine the "construct validity" of an assessment by examining the consistency of student performance across tasks. But again, such analyses simply seek to ensure that the test is true—that it's really measuring what it says it is measuring.

What We Are Testing

As science teachers, we already know the formula for better assessments. First, define *exactly* what you want to assess: the learning outcome. Then, choose a method (assessment) that measures *that variable* and no others. Having one variable at a time makes a good experiment, a good science class, and a great assessment. While we can't eliminate sources of variance in student performance (distractions, motivation, prior experience, prerequisite skills), we can strive to understand them as we interpret assessments.

Knowing the formula for better assessments and being able to apply it are two very different tasks—in Bloom's taxonomy and in the classroom. Consider this scenario:

Jim and Maria are veteran teachers in a suburban

school. Both teach units on photosynthesis to their general biology students. Both classes do hands-on experiments, read and respond to study guides, listen to explanations, answer higher-level questions, and review in groups and with audiovisual aids. When it is time for the test, Jim's classes get a long matching test, while Maria's test has more text and multiple-choice questions. Both classes' scores fall roughly on a "bell curve" distribution.

Maria and Jim frequently look for ways to take small steps toward reform. In the lounge they ask, "Which test format is better?" After deciding to re-test their classes with one another's exams, once again they find a spread in student grades. But this time there is an important difference—students who did well on one format did not necessarily do well on the other format!

When Jim and Maria turn to their administrator (a former biology teacher) for help, she suggests a third format. Students in the two classes are each given a blank sheet of paper and a small Coleus plant. The teachers say, "Show how this colorful organism gets food." First, students react with dismay. Then most get to work. Some draw chloroplasts, some write prose, some sketch flow charts.

Jim and Maria decide on a simple rubric to determine the scores: Those who relate the "food" process to photosynthesis are scored correct no matter how they express it. Among the wrong answers are essays talking about "plant food in the soil," lists of disconnected terms, and diagrams of plants with long roots. Only about half of those whose assessments end up in the correct pile had scored well on the objective tests!

For Maria, Jim, and tens of thousands of us, the question of what we are testing is a constant concern. We don't find it difficult to generate grades. In fact, we often get a bell curve on an easy test but without confidence that any learning has occurred at all.

It's difficult to develop assessments that don't confuse the variables. Traditional objective tests do. Much of the variance we see in multiple-choice test scores is caused by students' ability to read. Rote association produces most of the success students have on matching tests; they can memorize, but can they understand? Essays may measure understanding; but to be successful, students have to have some writing skills. Ideally, an *authentic* assessment would assign grades that correspond almost entirely to a student's achievement of a single *learning outcome* of the unit and not some other outcome, such as reading or writing ability. (That's not to say that critical reading isn't vital; it should be taught and tested, too.)

To reach the vision of the Standards, we will need to understand, use, and explain new forms of assessment. Changes will begin with classroom "tests" but must extend through the district, state, and national levels. This brief primer is designed to help teachers and schools begin to explore new ways to measure their progress toward the Standards.

An Assessment Primer

We often use the terms assessment and evaluation interchangeably, but they are not the same. Assessment provides a general picture of what students know and are able to do. We use assessments to improve programs, to measure the effect of curriculum changes, and to compare programs and populations.

When this data is quantified into a grade and attached to a student, it becomes an evaluation. We use evaluations to compare and rank students' abilities in science and to provide an accountability mechanism for us and our communities. In-class evaluations are most important to us because they help us direct instruction. Those that result in grades have the greatest meaning to students and parents. Often, evaluations that determine college entrance drive program and system changes.

Today, from classroom through district, state, and national levels, we realize that evaluation must change to fit what students need to learn and how they are able to learn it. Standard A stresses that all assessments should correlate well with the intended purposes of instruction. Therefore, we must first agree on a limited number of specific outcomes for each instructional unit. (The Content Standards provide a good place to begin.)

Once we have the outcomes clearly in focus, Standard B asks us to use achievement data wisely—not only to measure student performance, but also to measure the instructional environment in which students learn. Progress can only be meaningful when we compare it to the students' *opportunity to learn.*

As we pace instruction, information on what isn't working is as important as information on what is. In the scenario above, why were students who could draw or write about photosynthesis unable to succeed on a multiple-choice test? Was it their reading level? Were there clues in the wording of the items? Or were there hidden meanings? As we move toward the Standards in assessment, we must always use the results of tests as mirrors of instruction.

Because the data generated by assessments leads to decisions and actions, Standard C asks us to ensure that if we were to repeat the same test or a similar one, the results would be the same (that tests are *reliable).* But to get better reliability from our assessments, we must first ensure that they actually measure what they claim to measure (that the tests are *valid).* Tests that are closely tied to objectives are most likely to convey accurate information each time they are given.

Standard D challenges us to identify and eliminate inaccuracies that can infect testing. The informal evaluations we constantly use as teachers are filtered through a wealth of personal understanding and experience. But because tests also send high-profile messages to a public that may lack such background, we must carefully examine the messages test scores convey. A biased test that is highly reliable (providing the same results each time it is given) may give a message that is completely invalid.

Finally, Standard E asks our profession for a sound dose of humility. Any single assessment has limitations, and none is totally free from contamination from unwanted sources of variance. When we draw inferences from data—whether they are grades, generalizations, or assessments of instruction or program effectiveness—we must always be mindful of the weaknesses that are inherent in any single method or group of methods of assessment.

On the Firing Line

In assessment more than any other area, moving toward the Standards may put us on the defensive:
- "How can you justify

giving my child a B? He always got As on Mrs. Sampson's tests."
- "What you are doing to help my child do well on the College Entrance Exam?"
- "Mr. Beckman, some students are asking to drop your class because they think your tests are too hard."

Even if we really want to change assessment, we might be reluctant to face the consequences. While we may experiment with authentic assessments in daily lesson plans, we might resort to traditional tests when giving grades. In districts that publicize college entrance scores, it's not uncommon to find classroom teachers waiting for "guidance" on what to teach from state or national testing services.

Despite great intentions, many of us may feel caught between the Assessment Standards and external factors that determine so much about our students' futures. In many states and local communities the big tests have become "the tail that wags these dog," determining what type of instruction takes place. (Even the Content Standards themselves are divided into grade groupings—K–4, 5–8, and 9–12—because that is how major state assessment tests have been organized.)

The good news is that in recent years, many major assessment efforts have begun to exert a powerful influence on curriculum. Although few are incorporating true performance tests at present, many are moving beyond "inert" knowledge or facts. If you feel reform efforts are moving far too slowly, consider these scenarios in which administrators and test designers find themselves:

"The legislature wants a state test and will be looking at the results. We know that we should move to authentic performance-based measures. But if we produce one that is far ahead of current instruction, no one will pass. Imagine the bad press. So we'll wait...."

"Colleges want entrance examinations that predict which students will succeed in 100-level courses, which are even more verbally oriented than most high school courses. So we can't design an entrance test that will measure success on the job or in research; we need items with predictive ability for the freshman year of college. The tests can't change until the courses do."

So who moves first? Playing chicken on assessment won't help. The Standards demand a partnership between precollege and higher education.

Endnote: Teacher Assessment

Although the Standards do not mention teacher evaluation, the implications for job performance assessment are inescapable. Moving toward the Standards will be a highly complex process. Not every path will be smooth. When we misstep, we must be confident that we can pick ourselves up and try again without fear of being criticized. We need to be able to trust that our supervisors will understand and support this journey and be partners on it.

It's been only a few years since our profession decried formula teacher evaluation. When "checklist" or "recipe" instructional systems gained widespread popularity among principals, science teachers' eloquent responses helped evaluators understand that such formulas may not be appropriate in inquiry-based classrooms. We will need to voice a similar message about the Standards to those who assess us. Practically speaking, the responsibility for communicating classroom indicators of reform will again fall to us as teachers. We will have to say to our evaluators:

- Good science instruction is complex, difficult, and seldom "neat." When we urge students to be skeptical and curious, classrooms will become more active.
- In good science instruction, experiences follow student interest and challenge student misconceptions. Some students, often those who got "easy As" in traditional classrooms, will feel uncomfortable.
- Good science teachers seldom finish a "lesson" in a single class period. Great science teachers send their students out of the room with more questions than answers.
- Tests can only partially assess the outcomes of Standards-based science instruction. We must consider using many forms of authentic evaluation.
- Sound Standards-based instruction takes time. The pace of content coverage is not a valid assessment of teacher competency.

We can help ourselves and our profession by conveying the messages in the Assessment Standards to our supervisors and communities in terms that are readily understood:

- Good teachers assess their students continually, not just at test time.
- Assessment helps us pace instruction and gauge the opportunities students have for learning, as well as their ability to take advantage of these opportunities.
- Authentic assessments take time and may look very different from the ones students and parents are used to.
- While good assessments (and good science instruction) may seem difficult, there is constant effort to ensure that they are fair and unbiased.
- Finally, the information from any one assessment is only one piece of the puzzle that is instruction.

We cannot travel the path toward better instruction without the map provided by the five Assessment Standards:

- coordination with intended purposes: authentic assessments
- measuring student achievement and opportunity to learn
- matching technical quality of data with consequences
- avoiding bias
- making sound inferences

The discussions that follow offer several routes that many teachers have found helpful. While the journey you'll travel with your students toward new assessments may be unique, you won't be traveling alone.

About Assessment Standard A

Coordination with Intended Purposes

Designing an assessment is an intricate and difficult task; the resulting data drives curriculum and influences decisions made about students. It is essential that every evaluation tool be crafted so that educators can accurately measure clearly stated outcomes.

Reflecting Real Achievement

"Mirror, mirror on the wall..." Assessments reflect the instructional process, but do they show us the truth, or do they show us only what we want to see? When assessments are designed mainly to be convenient or to generate easily analyzed data, they may give distorted images of the outcomes we are trying to measure. Hacket and Floore summarize the problem by explaining, "What you test is what you get." They also point out, "Assessment does not evaluate whole realms of performance. Traditional tests... deny major accomplishments [of]...auditory and kinesthetic learners...social processes and skills..."

The search for new mirrors—new assessments—to reflect what we do has produced numerous alternative assessments, performance assessments, holistic assessments, and outcome-based assessments—all with the stated purpose of measuring a specified outcome and *only* that outcome. All these variations can be grouped under the general term "authentic assessment." Hart (1994) defines it this way:

> *"An assessment is authentic when it involves students in tasks which are worthwhile, significant, and meaningful....[They] look and feel like learning activities, not traditional tests...they communicate to students what it means to do their work well by making explicit the standards by which that work will be judged."*

By linking instruction and assessment, those who develop assessment instruments assume that the tasks will have greater validity than those of an assessment that "changes the rules" by asking students to apply what they did in class to paper-and-pencil questions.

Authentic (or performance) tasks are certainly not the complete solution to the problem of assessment. What

---

ASSESSMENT STANDARD A

Assessments must be consistent with the decisions they are designed to inform.
- Assessments are deliberately designed.
- Assessments have explicitly stated purposes.
- The relationship between the decisions and the data is clear.
- Assessment procedures are internally consistent.

Reprinted with permission from the *National Science Education Standards*. © 1996 National Academy of Sciences. Courtesy of the National Academy Press, Washington, D.C.

---

these activities gain in validity is often offset by losses in reliability, meaning that they may accurately measure a learning outcome but not consistently over time or from student to student. Also, administering a performance task to more than one student at a time can lead to problems of discipline, concentration, or control of other environmental variables. Nevertheless, performance tasks have produced statistically defensible data in large-scale administrations and are now part of many statewide assessment programs.

How can we begin to implement hands-on performance evaluations? Based on Doran and Hejaily (1992), here are 13 steps:

1. Select the program that you want to evaluate.
2. Choose a key skill outcome in this program.
3. Select the content area in which the skill will be evaluated.
4. Avoid developing a performance task for a skill that is easily evaluated on paper.
5. Write a behavioral objective.
6. List the materials that will be needed.
7. Determine any necessary safety precautions.
8. Develop questions for each task, and be sure they are not ambiguous.
9. Write directions that are clear and concise.
10. Use graphics to illustrate the setup, directions, etc.
11. Check the reading level of the directions, and make adjustments as necessary.
12. Develop scoring procedures using a complete, concise rubric to ensure reliability.
13. Conduct a trial of the assessment. Begin with a small number of students; then gradually broaden the range.

If we want to see the results of our instruction through the mirror of assessment, our evaluative tools should look as much like our instruction as possible. Every time we change the context (from lab to paper-and-pencil, for example), we create distortions in the mirror that will hamper decisions we make based on the assessment results. Deliberate design of better assessments won't be easy, but there is a wealth of resources in our profession to help us in the effort.

Resources for the Road

Association for Supervision and Curriculum Development (ASCD). (1995). *Designing Performance Assessment Tasks.* Alexandria, VA: Author.

Barnes, Lehman W., and Barnes, Marianne B. (1991, March). Assessment, Practically Speaking. *Science and Children, 28* (6), 14–15.

Bell, Beverley. "Interviewing: A Technique for Assessing Science Knowledge." In Shawn M. Glynn and Renders Duit (Eds.), (1995). *Learning Science in the Schools: Research Reforming Practice.* Mahwah, NJ: Lawrence Erlbaum.

Bergman, Abby Bary. (1993, February). Performance Assessment for Early Childhood. *Science and Children, 30* (5), 20–22.

Comfort, Kathleen B. (1994, October). Authentic Assessment: A Systemic Approach in California. *Science and Children 32* (2), 42–43, 65–66.

Doran, Rodney L., Boorman, Joan, Chan, Alfred, and Hejaily, Nicholas. (1992, April). Successful Laboratory Assessment. *The Science Teacher, 59* (4), 22–27.

Doran, Rodney L., Boorman, Joan, Chan, Fred, and Hejaily, Nicholas. (1993, September). Authentic Assessment. *The Science Teacher, 60* (6), 37–41.

Doran, Rodney, Chan, Fred, and Tamir, Pinchas. (1998). Science Educator's Guide to Assessment. Arlington, Va: National Science Teachers Association.

continued next page

Finson, Kevin D., and Beaver, John B. (1994, September). Performance Assessment: Getting Started. *Science Scope, 18* (1), 44–49.

Hart, Diane. (1994). *Authentic Assessment: A Handbook for Educators.* Menlo Park, CA: Addison-Wesley.

Herman, Joan L., Aschbacher, Pamela R., and Winters, Lynn. (1992). *A Practical Guide to Alternative Assessment.* Alexandria, VA: Association for Supervision and Curriculum Development (ASCD).

Kleinheider, Janet K. (1996, January). Assessment Matters. *Science and Children, 33* (4), 23–25, 41.

LeBuffe, James R. (1993, September). Performance Assessment. *The Science Teacher, 60* (6), 46–48.

Liftig, Inez Fugate, Liftig, Bob, and Eaker, Karen. (1992, March). Making Assessment Work: What Teachers Should Know Before They Try It. *Science Scope, 15* (6), 4, 6, 8.

McMahon, Maureen M., and Yocum, Charles. (1994, October). Video Quizzes: An Alternative Assessment. *Science and Children, 32* (2), 18–20.

Meng, Elizabeth, and Doran, Rodney L. (1990, September). What Research Says…About Appropriate Methods of Assessment. *Science and Children, 28* (1), 42–45.

Price, Sabra, and Hein, George E. (1994, October). Scoring Active Assessments. *Science and Children, 32* (2), 26–29.

Radford, David L., Ramsey, Linda L., and Deese, William C. (1995, October). Demonstration Assessment. *The Science Teacher, 62* (7), 52–55.

Reichel, Anne Grall. (1994, October). Performance Assessment: Five Practical Approaches. *Science and Children, 32* (2), 21–25.

Tetenbaum, Zelda. (1992, March). An Ordered Approach. *Science Scope, 15* (6), 12, 14, 18.

Tippins, Deborah J., and Dana, Nancy Fichtman. (1992, March). Culturally Relevant Alternative Assessment. *Science Scope, 15* (6), 50–53.

Treagust, David F. (1995). Diagnostic Assessment of Students' Science Knowledge. In Shawm M. Glynn and Renders Duit (Eds.), *Learning Science in the Schools: Research Reforming Practice.* Mahwah, NJ: Lawrence Erlbaum.

About Assessment Standard B

Measuring Student Achievement and Opportunity To Learn

As teachers measure student achievement, they must always compare their results to the standard of "opportunity to learn." Test scores can reflect not only what students know, but also teachers' knowledge and skills, the degree of coordination in the program, the equipment and environment available for student experiences, and the community support behind student growth.

Credit Where Credit Is Due

No student learns in isolation, and no assessment should be used without considering the context within which it was designed. Students achieve because they have constructed their own science understandings, but they do that only when they are in an environment that challenges their prior knowledge and pushes them forward. Thus, we have to look at *every* set of test data not only for what it tells us about learners, but also for what it tells us about the *opportunities* we have provided *for* learners.

More and more frequently, media are publishing "school report cards." So, it falls to us as teachers to be proactive. We can help our communities assess and value what is really good in their schools, and we can help them consider what support they are offering their schools.

In states where comparisons of test scores among districts make front-page headlines, it is common to find that districts with lower scores also have lower-quality facilities and lower per-capita support for education. What does that imply? It is hard to separate the variables of low family income and low school funding because they occur so often together. But in laboratory schools and controlled situations, there is strong empirical data to support the fact that proper equipment and space, low class sizes, and good access to references all increase students' "opportunity to learn" and, in turn, their test scores.

When Jaeger, Gorney, and Johnson (1994) analyzed "school report cards" across the nation, they found that communities most often valued information on standardized test scores, student en-

> **ASSESSMENT STANDARD B**
>
> Achievement and opportunity to learn science must be assessed.
> - Achievement data collected focus on the science content that is most important for students to learn.
> - Opportunity-to-learn data collected focus on the most powerful indicators.
> - Equal attention must be given to the assessment of opportunity to learn and to the assessment of student achievement.
>
> Reprinted with permission from the *National Science Education Standards.* © 1996 National Academy of Sciences. Courtesy of the National Academy Press, Washington, D.C.

gagement (how much they liked school), school success, school environment, level of staffing, teacher characteristics, program offerings, school facilities, student services, background characteristics of students, and school finances. Test scores played only a small role. But without constant reminders from us, the public often grabs on to test scores as the part of the report card that is easiest to understand and use for comparisons.

NSTA offers an alternative system for assessing science programs, one that asks schools to rate various parameters, comparing what schools are doing in each area to what teachers, administrators, students, parents, and other community members think they should be doing. These NSTA *Guidelines for Self-Assessment* provide a unique sort of report card, since students will never have the opportunity to learn something that isn't valued to begin with. There are 70 areas in the secondary *Self-Assessment* report card. Here are some examples from the section on assessment:

52. District-wide criterion-referenced tests developed by teacher committees from the district are used....
53. External examinations and standardized tests are used to assess individual student and group achievement....
54. Planned, conscientious efforts are made to assess student progress and achievement in relation to objectives beyond knowledge and recall of information.
55. Our science curriculum's general effectiveness is judged by our students' performance on nationally normed examinations....
56. Attempts are made to assess student attitudes....
57. A formal evaluation of curriculum has been conducted within the past seven years by an outside consultant....
58. Student achievement is assessed in terms of success in meeting previously established performance objectives.
59. Teachers make anecdotal records of individual students' outstanding or special activities, accomplishments, problems, and behaviors in science.
60. Science enrollment statistics are gathered and examined to determine if any identifiable groups of students are not well served by the curriculum.
61. A variety of means, in addition to report cards, are employed to communicate with parents....
62. Diagnosis of student progress in science is an ongoing part of evaluation.
63. Each science course meets...for the recommended amount of time for that course.
64. Care is taken to minimize the effects of cultural bias of instruments (and teachers), reading difficulties, and physical handicaps.
65. Students are involved in evaluating curriculum, courses, and activities.
66. Students are involved in self-evaluating their progress.
67. Each science course uses a practicum [practical/performance assessment] to evaluate....
68. [A]n organized attempt is made to identify students

who appear to not be well served by the program....
69. When students begin a course, they know the minimum requirements that must be met for passing that course.
70. The science staff members have visited other schools to establish comparison standards.

The NSTA assessment instrument doesn't imply that these are *always* the right parameters to measure good programs and good evaluation systems; it simply asks teachers and others what *they* believe is important and then measures a program's progress in comparison to those parameters.

Just as student scores should be seen in light of student opportunity to learn, a schools' score can be seen in light of the school's "opportunity" to achieve. If students, teachers, administrators, or schools don't value an outcome, it will never happen.

When grades are high—for students, teachers, or schools—it's easy to share credit. But when achievement isn't so high, it's often just as easy to point fingers of blame. But the Standards remind us that "it takes a whole community to educate a child"—and about that many to pass a test! So we might use the Standards as a starting point to evaluate and to improve *everyone's* opportunity to learn and move toward the vision offered by the Standards.

Resources for the Road

Jaeger, Richard M., Gorney, Barbara E., and Johnson, Robert L. (1994, October). The Other Kind of Report Card: When Schools Are Graded. *Educational Leadership, 52* (2), 42–45.

National Science Teachers Association. (1989). *Guidelines for Self-Assessment: High School Science Programs.* Arlington, VA: Author.

National Science Teachers Association. (1989). *Guidelines for Self-Assessment: Middle/Junior High School Science Programs.* Arlington, VA: Author.

National Science Teachers Association. (1989). *Guidelines for Self-Assessment: Elementary School Science Programs.* Arlington, VA: Author.

About Assessment Standard C

Matching Technical Quality of Data with Consequences

Assessment data fuels decisions on students, teachers, programs, and systems. To move toward the Standards, the profession must have confidence in the technical quality of that data. Assessments must be valid, authentic, and reliable. Decisionmakers must have data in which the confidence level is consistent with the consequences of the decision to be made.

Passing the Test

"Ladies and gentlemen, place your bets!" calls the dealer. Those who understand the rules of the game dig deep into their pockets, while those new to the table wager more carefully. But in games of chance, knowledge is often deceptive, and even the card counters are often fooled.

The education of our nation's youth should never be a game of chance. The decisions we make and the paths we choose should be guided by reliable information about the potential effects of those decisions on students. To get that sort of data, we must use assessments that are both reliable and valid.

Reliability—making an assessment that can be counted on to give similar results when it is repeated many times—isn't hard to achieve. In fact, the methods used in most classrooms are so reliable that they create false confidence and generate bad decisions. Here are two examples:

Mrs. Wong uses multiple-choice tests for each chapter assessment in physical science. The questions have long stems (50 words or more), and students are asked to complete 50 questions in a single class period. Since Mary consistently scores higher than any other student on these tests, Mrs. Wong recommends Mary for Advanced Placement Chemistry, a lab-based program. Mary flounders in the course.

Hennington Community College enrolls students in its child care education program on the basis of essays they write. Most of the core courses use the essay format for exams. But the results of a recent survey show that Hennington's graduates seldom remain in the child care field.

ASSESSMENT STANDARD C

The technical quality of the data collected is well matched to the decisions and actions taken on the basis of their interpretation.
- The feature that is claimed to be measured is actually measured.
- Assessment tasks are authentic.
- An individual student's performance is similar on two or more tasks that claim to measure the same aspect of student achievement.
- Students have adequate opportunity to demonstrate their achievements.
- Assessment tasks and methods of presenting them provide data that are sufficiently stable to lead to the same decisions if used at different times.

Reprinted with permission from the *National Science Education Standards.* © 1996 National Academy of Sciences. Courtesy of the National Academy Press, Washington, D.C.

Both situations describe assessments that measure *something* very reliably, but what is it? It may not be the variable the instructor identified as a learning goal. Mrs. Wong's test is more likely measuring reading ability, not science understanding. And Hennington's curriculum may measure abilities in writing, organizing, or reading, but certainly not verbal communication or child management outcomes.

Teachers bet their lesson plans on assessment data, colleges bet their admissions policies on standardized test scores, and students often bet their entire working lives on what test scores tell them about themselves.

Test data can be highly reliable—giving about the same results every time it is used—but have low content validity (relationship to *what* is taught) or low predictive validity (relationship to *future success*). What we *think* the scores are saying and what they are *really* telling us can be very different. To move toward the Standards, we will have to learn far more about the various factors that affect students' classroom test scores.

Let's consider a few of them:
Cognitive Level. Piaget told us that students at the concrete operational level of development cannot reason in words. Today we know that theory is almost always true. So the unpredictability of early adolescents' ability to translate what they know into words is always a source of variance in their test scores.

Individual performance testing is a possible but unrealistic option for frequent use in typical classrooms. Yet research tells us that *any practical component* that we can add to a test increases its validity. So we might try doing a demonstration and asking students to respond to questions on paper. We might leave the apparatus used in labs on the table during a test to prompt visual memory. Or we might consider laserdisc or CD-ROM displays of real-life situations to help students build bridges from the concrete to the formal in a test.

Emotion. Every one of us knows a student who is "test phobic." This fear can drive learning, attention, and memory. The cognitive networking that allows a student to associate what she or he knows with what the question is asking may be impossible in the presence of fear.

Tests that begin with easy questions can relax students. Physical activity and social interaction can ease emotional blocks. Even humor can improve test performance. (Try a cartoon or use school situations as real-life stem scenarios.) Allowing a fruit drink or milk during the test can help. Even the simple technique of inviting students to talk about their fears can reduce the variance caused by test phobia.

Reading Ability. The ability of students to construct meaning from informational text is always a factor that limits the validity of a written test. We can help reduce this effect by keeping stems short and adding diagrams. For students who need a major modification, consider asking senior citizen or other community volunteers to audiotape tests. Always use clear type (no dot matrix or dittos), plenty of white space, and short sentences on written tests. And don't hesitate to *teach* students how to read tests. It's a learned skill.

Test-savvy students can capitalize on their reading ability. They quickly learn to recognize context clues to the right answer if we aren't careful to avoid leaving clues that can net a deceptively high score. Clues can lurk in the stems of multiple-choice questions and in choices that are grammatically parallel when others are not. Be aware that the right answer is often the longest one. Finally, it is *very* difficult to produce a valid matching test that measures anything but rote memory or word recognition.

Listening. The least taught of the language arts may be one of the most important for test success. Poor listeners miss directions and environmental clues, and emotions may make weak listeners even less able to comprehend directions at test time. While it may seem sensible to deliver oral directions or questions for a test (to avoid reading problems), this technique may create even greater problems because it biases the test toward students who are good listeners and those who can respond at the pace of the questioner.

Physical Distractions. We recognize that many students have attention deficit disorders. Even more have problems maintaining concentration in high-stakes situations. Noise, clutter, movement, and even the temperature of the classroom can produce dramatic changes in achievement for many students.

To get good data, experienced teachers plan the environment for testing as carefully as they plan the questions. Then they demand the respect of others in the school community to minimize or keep out distractions, announcements, and physical changes that can lower scores. At the same time, we also recognize that some students perform better if we relax normal classroom rules a bit. A corner study carrel may be ideal for one student, while another student may need a mid-test stretch.

Good teachers know there is only so much they can do to create authentic tests and limit the sources of error in their data. We can minimize risk by using a wide variety of measures and never relying on a single source of information. By carefully administering multiple assessments, we can make decisions that are likely to help our students become winners in school, career, and life.

Resources for the Road

Hardy, Garry R., Sudweeks, Richard R., Tolman, Marvin N., Tolman, Richard R., and Baird, J. Hugh. (1991, October). Does Listening Ability Affect Test Scores? *Science and Children, 29* (2), 43–45.

Marshall, Gail. (1991). Evaluation of Student Progress. In David Holdzkom and Pamela Lutz (Eds.), *Research Within Reach: Science Education.* Charleston, WV: Appalachia Educational Laboratory.

McMahon, Maureen M., and Yocum, Charles A. (1994, October). Video Quizzes: An Alternative Assessment. *Science and Children, 32* (2), 18–20.

Radford, David L., Ramsey, Linda L., and Deese, William C. (1995, October). Demonstration Assessment. *The Science Teacher, 62* (7), 52–55.

Shick, Jacqueline. (1990, September). Textbook Tests: The Right Formula? *The Science Teacher, 57* (6), 33–39.

Sylwester, Robert. (1994, October). How Emotions Affect Learning. *Educational Leadership, 52* (2), 60–65.

Tolman, Marvin N., Sudweeks, Richard, Baird, J. Hugh, and Tolman, Richard R. (1991, September). Does Reading Ability Affect Science Test Scores? *Science and Children, 29* (1), 44–47.

Wise, Kevin C. (1993, September). New Teacher Feature: Testing Tips. *Science Scope, 17* (1), 51–52.

About Assessment Standard D

Avoiding Bias

In assessment as in teaching, care must always be taken to avoid stereotypes and bias. Just as the styles of learners differ, their perspectives may vary based on their background, environment, or areas of ability. Assessment practices must accommodate diversity, and assessment data must always be examined for signs of bias.

The Road Less Traveled

At its core, learning is the most personal of acts. Each learner compares his or her instructional environment to a personal world created by environment, culture, ability, and prior learning. While the destination of each journey may be defined by the curriculum, the starting point for every student is unique.

Like teaching, assessment must accommodate the individual differences of each student. In measuring progress along the road, we must consider where students begin and what their style of travel is. One learner may take a direct route from a readily identified starting point; another may follow a circuitous path, stopping to smell the flowers as he or she travels. Each is learning, and we must value each step through the assessments we choose.

There is a wealth of data on cultural bias in testing. Variations in language, diagrams, and scenarios that evoke different stereotypes can all result in deceptive test data. There is less research on the bias toward learning styles in various forms of tests, so we have very few resources to help us modify testing for students with learning disabilities in the general education classroom.

While we as educators recognize that learning is a personal journey, there is far less consensus among parents and the general public that these differences should entitle students to take different tests—and to earn the same grade for making equal progress toward different end points.

The Portfolio Option

Many of us strike a compromise between individualizing testing and relying on arbitrary standards by working with students to develop *portfolios*. Named from the

ASSESSMENT STANDARD D

Assessment practices must be fair.

- Assessment tasks must be reviewed for the use of stereotypes, for assumptions that reflect the perspectives or experiences of a particular group, for language that might be offensive to a particular group, and for other features that might distract students from the intended task.
- Large-scale assessments must use statistical techniques to identify potential bias among subgroups.
- Assessment tasks must be appropriately modified to accommodate the needs of students with physical disabilities, learning disabilities, or limited English proficiency.
- Assessment tasks must be set in a variety of con-

continued next page

> texts, be engaging to students with different interests and experiences, and must not assume the perspective or experience of a particular gender, racial, or ethnic group.
>
> Reprinted with permission from the *National Science Education Standards.* © 1996 National Academy of Sciences. Courtesy of the National Academy Press, Washington, D.C.

leather-bound collections of artists and writers, the portfolio of a learner charts his or her personal journey toward achievement. Because portfolios start wherever learning begins, they show an individual's progress over time.

As an assessment tool, portfolios provide a way to avoid cultural and environmental bias as well as to individualize assessment for challenged learners. Like the professional's collection, student portfolios can be presented to the next level (another course, college, or job) as a supplement to grades, offering additional evidence of achievement.

In actual use, a portfolio can contain anything you or your students choose to include. But from a professional standpoint, the collection is most useful when it demonstrates linear achievement toward one or a small number of related learning outcomes. With clear outcomes in mind, you can develop a *rubric*, or achievement scale, *before* students begin, and thus simplify the grading process at the end. Students can help identify the purpose of the portfolio and can also contribute to the preparatory discussions about its contents:
- What will be required?
- How will pieces be selected?
- Will the portfolio contain only my best (polished) work? Or will it include work in progress?
- Where will the portfolio be stored?
- How much material will be kept there? (How many pieces? Gathered daily or weekly? For how many months?)

When students give input on these questions, the portfolio becomes more meaningful to them as a motivator and self-assessment tool. The value of the collection to external evaluators can become secondary to its value to the student.

Portfolios have taken widely different forms. Collections can track independent research; development of concepts in the form of writing, drawing, or diagrams; or self-evaluation. In 1992, the developers of California's Golden State Examinations (GSE) used the following three types of work samples in a pilot study:
- a problem-solving investigation that required students to design and conduct a research project
- a creative expression of a scientific concept using art, poetry, video, or music
- a writing sample that demonstrated growth in understanding of a scientific concept

In the GSE portfolio, students evaluated their own work using self-reflection sheets. Roughly 500 teachers and 4,000 students participated in the study.

Researchers have already drawn some preliminary conclusions from that study, which included objective forms of assessment as well. In biology, female students did better on the portfolio than on the objective section of the examination, in which males did better. In general, portfolio scores correlated more closely with grades in school than scores on the objective examination questions did. Differences based on ethnicity were minimal in the portfolios but significant in the objective tests. There was some correlation between portfolio scores, multiple-choice items, and open-ended items, and between portfolio scores and laboratory assessments, but the correspondence was not perfect.

Advocates for portfolios say that they "are more likely to elicit the true capability of most students, not just those motivated to do well on...one-shot tests" (Herman, Aschbacher, and Winters, 1992). But another researcher warns, "Using portfolios for external accountability may destroy their greatest benefit for teachers and students" (Case, 1994). When students know that the contents of their collection count toward grades, they often discard unfinished pieces or those that reflect prior misconceptions. In doing so, they may forget their own personal steps toward achievement and thus may be more likely to take the same missteps in the future.

As teachers, we are also moving to portfolios to track our own achievements in guiding student learning. Keeping evidence of student errors and misconceptions along with the old plan books can help us from repeating less effective methods next year.

Documenting student progress also provides excellent fodder for discussions with administrators about why final test scores are lower in a given year. Classes differ, so their end points can't easily be compared. With portfolios, we can track individual growth independent of classroom environment, cultural bias, learning style, or ability students brought with them to class.

Just as learning is personal, so is teaching. Every professional needs reassurance that students are progressing. For teachers and students alike, portfolios offer new ways to share achievement.

Resources for the Road

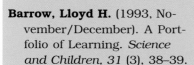

Barrow, Lloyd H. (1993, November/December). A Portfolio of Learning. *Science and Children, 31* (3), 38–39.

Bonnstetter, Ronald J. (1992, March). Where Can Teachers Go for More Information on Portfolios? *Science Scope, 15* (6), 28.

Case, Susan H. (1994, October). Will Mandating Portfolios Undermine Their Value? *Educational Leadership, 52* (2), 46–47.

Collins, Angelo. (1992, March). Portfolios: Questions for Design. *Science Scope, 15* (6), 25–27.

Doran, Rodney, Chan, Fred, and Tamir, Pinchas. (1998) Science Educator's Guide to Assessment. Arlington, Va: National Science Teachers Association.

Gitomer, Drew H, and Duschl, Richard. (1995). Moving Toward a Portfolio Culture in Science Education." In Shawn M. Glynn and Renders Duit (Eds.), *Learning Science in the Schools: Research Reforming Practice.* Mahwah, NJ: Lawrence Erlbaum.

Gustafson, Chris. (1994, October). A Lesson from Stacy. *Educational Leadership, 52* (2), 22–23.

Hamm, Mary, and Adams, Dennis. (1991, May). Portfolio Assessment: It's Not Just for Artists Anymore. *The Science Teacher, 58* (5), 18–21.

Herman, Joan L., and Winters, Lynn. (1994, October). Portfolio Research: A Slim Collection. *Educational Leadership, 52* (2), 48–55.

Martin, Megan, Miller, George, and Delgado, Jane. (1995, January). Portfolio Performance: Research Results from California's Golden State Examinations Science Portfolio Project. *The Science Teacher, 62* (1), 50–54.

O'Neil, J. Peter. (1994, January). Portfolio Pointers. *Science Scope, 17* (4), 32.

Tippins, Deborah J., and Dana, Nancy Fichtman. (1992, March). Culturally Relevant Alternative Assessment. *Science Scope, 15* (6), 50–53.

U.S. Department of Education, Office of Educational Research and Improvement. (1993, November). *Student Portfolios: Classroom Uses.* Consumer Guide No. 8. Washington, DC: Author.

U.S. Department of Education, Office of Educational Research and Improvement. (1993, December). *Student Portfolios: Administrative Uses.* Consumer Guide No. 9. Washington, DC: Author.

About Assessment Standard E

Making Sound Inferences

Assessments influence the plans we make for students, courses, and programs. As teachers move toward the Standards, they will rely on data each step of the way. It is crucial that in each decision, teachers keep in mind the strengths, weaknesses, assumptions, and inaccuracies inherent in every assessment.

Making the Grade

Have you ever met a former student in a store and asked her what she remembered from your class? You secretly hope the answer will be some higher-order understanding of a concept or application, but you hear, "We had fun...and I only got a C." When the books are packed up, the desks are cleaned, and the papers are filed at the end of the year, the permanent result of most students' participation in most secondary classes is the grade, a single letter that sticks with them as honor or chastisement for a lifetime.

In some cases standardized test scores accompany the grade on the permanent record; less frequently, students take with them portfolios of work or research. But for most students, the sum total of their science experience is the grade. Employers, institutions of higher learning, and even successive generations make judgments about a year of work from single numbers or letters, with little or no understanding of the variables that went into their calculation.

We, too, use scores and grades to rate ourselves, our schools, and our systems. In many states student achievement scores are published and compared across districts. Colleges are rated by the average scores of entering freshmen. We may find ourselves rated, scheduled, or able or unable to "recruit" future students based on the grades we give.

For those who have studied the variables inherent in every assessment score, the emphasis on scores and grades seems much ado about nothing. The consequences are far out of proportion to the confidence we have in the accuracy of the data.

The Standards speak clearly for change: When we use scores and grades, we must keep in mind the assumptions on which they were based and the possibili-

> **ASSESSMENT STANDARD E**
>
>
>
> The inferences made from assessments about student achievement and opportunity to learn must be sound.
> - When making inferences from assessment data about student achievement and opportunity to learn science, explicit reference needs to be made to the assumptions on which the inferences are based.
>
> Reprinted with permission from the *National Science Education Standards.* © 1996 National Academy of Sciences. Courtesy of the National Academy Press, Washington, D.C.

ties of error in their determination. "Don't make more of grades and scores than they deserve," the specialists remind us.

But that's easier said than done. There is nothing like a skeptical parent at conferences, for example, to discourage our first steps toward

Standards-based assessment. Seeley (1994) describes the dilemma:

> [T]eachers are encouraged to use a number of types of alternative assessments to guide instruction and monitor student thinking. How can all this information be recorded in a single letter grade? Teachers are encouraged to challenge students to do complex tasks and to communicate effectively. And yet they realize that low grades may have a negative impact on these efforts.

Relying on Rubrics

For many activities, rubrics offer an answer. A rubric is a clear, concise statement, usually accompanied by a quantitative score, of the levels of achievement a student might demonstrate on a given task. In short, it is a way to convey in objective, simple language our understanding of the cognitive development of students.

Rubrics force us to communicate clearly because we can't even begin to construct a rubric until we first decide *exactly* what learning outcome we want students to achieve. Then we have to analyze the concepts or skills that will be required to achieve that outcome. To do this, we must know both the topic *and* the learner. (For an example of a rubric, see page 108. See also Resources for the Road, page 60.)

It can be helpful to use concept maps to diagram the components of a learning outcome before we begin to identify levels of achievement. We can then set the highest level of performance at the top and progressively lower levels further from the ideal, thus charting the path toward the goal. This process can aid our professional growth.

Rubrics give parents and students a clear, objective guide to evaluate performance and authentic curriculum-related tasks. But they serve many other purposes as well. They show students how to improve, since the next step up doesn't seem as far away as "perfection." Because rubrics can be converted into numbers, they can be averaged with traditional scores, letting us vary our grading criteria and move toward more authentic assessment one step at a time. Perhaps most important, rubrics encourage teachers and systems to analyze the cognitive steps we are asking students to take.

Rubrics usually net clear data that are perceived as fair and accurate. But when this does not happen, we can learn a lot. When the data from rubrics refuses to fall into neat patterns, we are forced to reassess our assumptions about how students are progressing or even whether they understand the outcome they are working toward.

Ideally, a rubric should trace a path toward a single skill or concept. When the outcome really requires two or more separate strands of growth, the path won't be linear, and there won't be an easy way to describe the learner at each step. That's when we need to step back, take a hard look at the task, and reassess our prior assumptions. If student growth doesn't come in definable steps, maybe we aren't measuring what we think we are.

It's never enough to tell a student, "Don't worry about your grade!" The number of schools and systems that have abandoned grading (without massive community backlash) is small. Most teachers, in most classrooms, will be grading students regularly throughout their careers. To move toward the Standards, we will have to making grading better. At the same time, we will have to educate students and the public about what grades really mean.

Resources for the Road

Berg, Craig A., and Clough, Michael. (1991, October). Generic Lesson Design: Does Science Fit the Mold? The Case Against. *The Science Teacher, 58* (7), 26–27, 29–31.

Hunter, Madeline. (1991, October). Generic Lesson Design: Does Science Fit the Mold? The Case For. *The Science Teacher, 58* (7), 26–28.

Jensen, Ken. (1995, May). Effective Rubric Design. *The Science Teacher, 62* (5), 34–37.

Kenney, Evelyn, and Perry, Suzanne. (1994, October). Talking with Parents About Performance-Based Report Cards. *Educational Leadership, 52* (2), 24–27.

Liu, Katherine. (1995, October). Rubrics Revisited. *The Science Teacher, 62* (7), 49–51.

Moran, Jeffrey B., and Boulter, William. (1992, March). Step-by-Step Scoring. *Science Scope, 15* (6), 46–47, 59.

Nott, Linda, Reeve, Colleen, and Reeve, Raymond. (1992, March). Scoring Rubrics: An Assessment Option. *Science Scope, 15* (6), 44–45.

Roth, Wolff-Michael. (1992, March). Dynamic Evaluation. *Science Scope, 15* (6), 37–40.

Seeley, Marcia M. (1994, October). The Mismatch Between Assessment and Grading. *Educational Leadership, 52* (2), 4–6.

Smith, Paul G. (1995, September). Reveling in Rubrics. *Science Scope, 19* (1), 34–36.

Changing Emphases

The National Science Education Standards envision change throughout the system. The Assessment Standards encompass the following changes in emphases:

LESS EMPHASIS ON	MORE EMPHASIS ON
Assessing what is easily measured	Assessing what is most highly valued
Assessing discrete knowledge	Assessing rich, well-structured knowledge
Assessing scientific knowledge	Assessing scientific understanding and reasoning
Assessing to learn what students do not know	Assessing to learn what students do understand
Assessing only achievement	Assessing achievement and opportunity to learn
End-of-term assessment by teachers	Students engaged in ongoing assessment of their work and that of others
Development of external assessments by measurement experts alone	Teachers involved in the development of external assessments

Reprinted with permission from the *National Science Education Standards*. © 1996 National Academy of Sciences. Courtesy of the National Academy Press, Washington, D.C.

Content Standards

*We must let go
of our preconception
that we must cover
all the topics in the
textbook.*

Mapping the Content Standards

There has been a groundswell of collaboration among science teachers, scientists, and researchers focusing on how students learn. Together, they have determined which concepts and themes are developmentally appropriate to the cognitive level of learners at various stages. The Standards reflect the result: away from presenting information to encouraging student discovery, and away from breadth of content to depth of coverage. Together we've recognized that in tomorrow's science classroom our role as teachers will be to provide an environment in which students are guided to become critical thinkers.

But the Standards do not focus solely on the methods through which science is learned. They define a solid core of essential content at every level. Although at first examination the Standards may seem to call for less content than traditional secondary courses, the scope of the content they call for is challenging. Perhaps most important, the Content Standards represent outcomes for *all* students to achieve.

Students enter science with a great many preconceived ideas about the natural world. By exploring and experiencing science, they must continuously examine and compare their preconceptions to the objective knowledge defined by the Content Standards. The path to critical thinking in science requires that students carefully analyze their premises as well as the flaws in their reasoning.

To achieve the Content Standards, the learning environment of a classroom must encourage students to take risks through open thinking, allow sufficient time for students to develop and test concepts, and promote cooperative learning and respect and equity in the sharing of ideas. Empowering students to create their own abstract conceptual models will take time. In the quest to provide an atmosphere to support students in their search for understanding, we must let go of our own preconception that we must cover all the topics in the textbook.

The Content Standards include not only outcomes in the traditional disciplines of life, Earth and space, and physical science, but also in the areas of
- Science as Inquiry, which delineates the essential abilities to do and understandings about scientific inquiry

- Science and Technology, which discusses what students should know about technological design and the relationship between science and technology
- Science in Personal and Social Perspectives, which covers such issues as personal and community health, natural resources, and environmental quality
- History and Nature of Science, which addresses the history and methodology of the sciences

The Content Standards present a distinctly traditional division among the sciences; one which, the authors of the document admit, may not completely reflect the nature of today's research or the perspective of many adolescent learners. While the division into life, Earth and space, and physical sciences may be somewhat arbitrary, the Standards encourage cross-disciplinary studies. The Unifying Concepts and Processes outlined in the Standards consider the following basic to all the sciences:

- Systems, Order, and Organization
- Evidence, Models, and Explanation
- Change, Constancy, and Measurement
- Evolution and Equilibrium
- Form and Function

While the Standards do not deal directly with programs of integrated science, they support these efforts by emphasizing that unifying concepts extend across programs and disciplines.

Because high school science courses are generally organized in and taught by specialists in life, Earth and space, or the physical sciences, the sections that follow are organized in that manner. Threaded through each area and each "Classroom in Action" vignette are examples of teaching and learning about the other Content Standards: Science As Inquiry, Science and Technology, Science in Personal and Social Perspectives, and History and Nature of Science. This approach demonstrates in a concrete way how these standards can be successfully integrated throughout every course.

This organization is primarily for the convenience of the reader and is not meant to suggest that courses and programs should be organized along traditional lines. How best to reach the Standards will be the decision of each system and each school across the nation.

Mapping the Content Standards

Science Content Standards

Content Standard: K–12

Unifying Concepts and Processes

STANDARD: As a result of activities in grades K–12, all students should develop understanding and abilities aligned with the following concepts and processes:
- systems, order, and organization
- evidence, models, and explanation
- constancy, change, and measurement
- evolution and equilibrium
- form and function

Content Standards: 9–12

Science As Inquiry

CONTENT STANDARD A: As a result of activities in grades 9–12, all students should develop
- abilities necessary to do scientific inquiry
- understandings about scientific inquiry

Physical Science

CONTENT STANDARD B: As a result of their activities in grades 9–12, all students should develop an understanding of
- structure of atoms
- structure and properties of matter
- chemical reactions
- motions and forces
- conservation of energy and increase in disorder
- interactions of energy and matter

Life Science

CONTENT STANDARD C: As a result of their activities in grades 9–12, all students should develop understanding of
- the cell
- molecular basis of heredity
- biological evolution
- interdependence of organisms
- matter, energy, and organization in living systems
- behavior of organisms

Earth and Space Science

CONTENT STANDARD D: As a result of their activities in grades 9–12, all students should develop an understanding of
- energy in the Earth system
- geochemical cycles
- origin and evolution of the Earth system
- origin and evolution of the universe

Science and Technology

CONTENT STANDARD E: As a result of activities in grades 9–12, all students should develop
- abilities of technological design
- understandings about science and technology

Science in Personal and Social Perspectives

CONTENT STANDARD F: As a result of activities in grades 9–12, all students should develop understanding of
- personal and community health
- population growth
- natural resources
- environmental quality
- natural and human-induced hazards
- science and technology in local, national, and global challenges

History and Nature of Science

CONTENT STANDARD G: As a result of activities in grades 9–12, all students should develop understanding of
- science as a human endeavor
- nature of scientific knowledge
- historical perspectives

Reprinted with permission from the *National Science Education Standards*. © 1996 National Academy of Sciences. Courtesy of the National Academy Press, Washington, D.C.

Physical Science

Change is occurring in physical science classrooms across the nation. Students are being offered alternative ways to build a knowledge base about the natural universe. Many traditional approaches to the teaching of physical science are being reexamined in the context of the Standards.

The essence of the change toward a more reflective physical science classroom is inquiry, which the Standards call "a basic and controlling principle" in science education. Inquiry is both an area for students to study and a method of teaching. The change toward inquiry-based classrooms will mean that physical science students will spend less time solving pages of similar mathematical problems and more time choosing the right method to solve a single problem.

The inquiry approach to teaching is one step beyond "science as process," because students not only observe, infer, and experiment, but they also combine these activities with scientific knowledge to reason and think critically as they develop understanding. Inquiry serves five purposes in a physical science classroom: to develop understanding of concepts, convey how we know what we know, develop understanding of the nature of science, develop the skills necessary to inquire independently, and develop the disposition to use science.

Once given the opportunity to use inquiry and the freedom to think and act on their own, students will ask questions, investigate using appropriate tools, think critically and logically, examine evidence and explanations, construct and analyze alternative explanations, and communicate arguments scientifically—and they will never again be satisfied with traditional, textbook-bound lessons.

To begin to foster inquiry, we need to discover how students think and what they know. To examine students' ideas, we can conduct interviews; participate in small-group discussions; and ask students to make drawings, diagrams, and concept maps. In all discussions, we should promote a respectful exchange of ideas so that students learn how to support one another.

Implementing an inquiry approach implies that student explanations will become a baseline for instruction. We cannot assume that students will progress at a specific rate. As they learn to work from their own ideas, they can move from guided investigation to less structured investigations.

Teaching through inquiry never implies that lessons end with unverified student ideas. By the time they reach high school, students have already experienced many concepts covered by the physical science standards. But we know that they also have many misconceptions, especially in key areas, such as conservation of matter and energy. A sound secondary physical science program begins with explorations that relate to earlier coursework and allows students to apply more mature logical reasoning to familiar contexts.

We must continually assess where students are and aid them in constructing accurate explanations in line with current scientific knowledge. The Standards define six core areas of content for physical science knowledge in secondary schools:

- *Structure of Atoms.* Matter is made up of atoms, which are themselves composed of even smaller components.
- *Structure and Properties of Matter.* Atoms and molecules interact with one another through bonding and forces.
- *Chemical Reactions.* Chemical reactions release or consume energy.
- *Motions and Forces.* Motion occurs when a net force is applied; gravitation, electricity, and magnetism are examples of forces.
- *Conservation of Energy and Increase in Disorder.* Energy is kinetic or potential; everything becomes less orderly over time.
- *Interactions of Energy and Matter.* Waves can transfer energy; electromagnetic waves include radio, microwaves, and infrared radiation.

Whether these content areas are integrated over three or four years, or self-contained in two or three years of physical science courses, they should be the focus of secondary physical science curricula. They are expectations for *all* students, unlike today when only some students take chemistry, and many never enroll in physics. In a Standards-based program, all secondary students will study physical science. Chemistry and physics will not be reserved for the college-bound and will generally occur earlier in the school program (in grade 9 or 10).

Real-world applications are essential to make physical science accessible to all students. Pilot programs in applied and active physics have been quite successful for students of varied abilities in many schools. The keys to success include limiting the entry-level mathematics needed (by including mathematics review in the program) and building on extensive exploration in every concept.

We must always be aware of prior content that students have mastered and the methods they've experienced so that we can meet students where they are. In each area of physical science, the secondary program must build on elementary and middle school programs.

As a foundation for the first three physical science standards, the Standards suggest that K–4 students become familiar with the properties of objects and materials (size, color, mass, temperature) and their ability to react with other substances. In grades 5–8, students should understand density, boiling point, and solubility. They should explore the nature of substances, compounds, and mixtures and the concept of conservation of mass in simple chemical reactions.

Building on this foundation, secondary students develop the ability to relate the macroscopic properties of matter to the microscopic structure. They study atomic theory, chemical formulas, and equations. Even at this level, however, it will be difficult for students to comprehend the particulate nature of chemical reactions, given the size and the number of particles involved. A study of the historical development of the concept of the atom can help students in their own conceptual growth. Hands-on experience with the macroscopic world of forces, motion, vibrations, waves, light, and electricity supports student understanding of the microstructure of matter.

The Standards encourage K–4 classrooms to begin the study of Motions and Forces by defining the location of objects with reference to a specified point, and then looking at the kinds of motion and the forces that control them. From this starting point, students in the middle grades study position, direction, speed, and the graphical representation of motion. Students gain an understanding of velocity and inertia, as well as balanced and unbalanced forces and their relationships to steady motion or acceleration. In grades 9–12, students examine the laws of motion, different types of forces, and electricity and magnetism.

The study of energy begins with experiments with light, heat, magnets, mirrors, and lenses at the K–4 level. Studying the production of heat by machines and chemical reactions and the conduction of heat by different materials is also appropriate at this level. In the middle grades, students consider different forms of energy and how energy is transformed from one form to another. In-depth investigations into the transmission, absorption, and scattering of light (including infrared and ultraviolet radiation) and the energy released by chemical reactions are appropriate topics. The study of electric circuits highlights the conversion of electrical energy into heat, light, sound, and chemical energy. These experiences enable students to investigate energy transformations quantitatively in grades 9–12 by measuring variables such as temperature change and kinetic energy.

Teaching physical science at the secondary level offers a wealth of opportunities for incorporating other aspects of the Content Standards: Science and Technology, Science in Personal and Social Perspectives, and the History and Nature of Science. For example, understanding the enterprise of science and its link with technology is an appropriate extension in many units.

In the same way, cost analysis, decisionmaking, global versus local resource use, environmental problems, human needs, population growth, the hazards of waste disposal, and personal and community health are all natural strands to consider when linking physical science to personal and social perspectives. Where integration does not occur within science courses themselves, we are challenged to define continuity across courses.

Science is a human endeavor that changes as we expand our knowledge base and gain better technologies for observing natural phenomena. Does nature wait patiently for us to discover scientific "truth," or do we grope toward understanding by systematically asking the right questions and designing tools to answer them? As we walk the pathways to implement the Standards, we might keep this question in mind, using inquiry to hone the sense of the scientific enterprise in our students and to heighten their awareness of its historical and cultural perspectives.

We as teachers can deal with the real pressures in the classroom and with the pressures that come with being agents of change. But taking the road less traveled—in small steps, one a time—can be rewarding for us and our students. Soon we'll find many of us are traveling that same road.

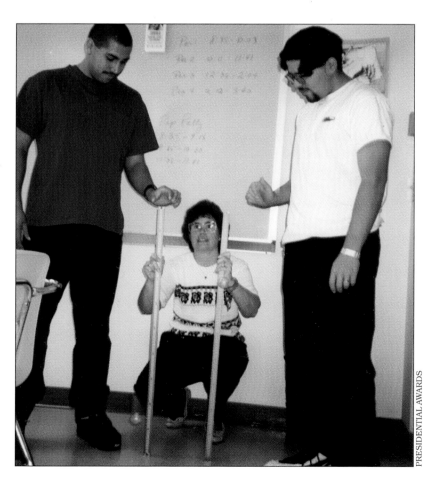

Physical Science 67

Structure of Atoms

Nature of the Learner

The study of atoms and molecules requires students to construct knowledge about particles they cannot see from indirect observations of models and data. While most students will be able to communicate their understandings of physical models, the ability to make inferences about atomic structure will be less common.

Chemistry also requires students to communicate their understandings in symbolic form (formal logic) and to solve problems using mathematics. Most students will require extensive practice to achieve these outcomes. When solving chemistry problems, students will prefer to rely on examples or algorithms, but this does not promote understanding. They will have much greater difficulty solving problems when they must choose their own method or apply an algorithm they already understand to a new situation. Initially, by focusing on what is happening chemically, students can develop conceptual models even without specific mathematical computations.

Physical models are appropriate for all students, but they may cause some students to overgeneralize and to miss the differences between the model and the real phenomenon. While mathematical models will be more difficult, they should be used when introducing students to a more formalized approach to modeling.

History and Nature of Science

Many historical discoveries in chemistry illustrate how the increased capacity of technology is directly related to advances in understanding. Because many concepts involve knowledge about particles students will never see, seeking foundations for "how we know" through review of classical chemistry investigations is very valuable. These investigations provide a historical perspective on model building as well as define the chain of evidence that leads us to express confidence in the atomic theory. Teaching with historical vignettes also enables us to avoid asking students to believe without seeing.

Nature of Instruction

A chemistry program with a broad spectrum of activities offering insight into qualitative and quantitative reactions is an ideal format to explore the nature of theory and to introduce current areas of uncertainty in science.

Demonstrations accompanied by Socratic questioning can be an effective technique. But we should be careful to avoid the types of demonstrations and labs that are flashy and create dramatic effects because often they pose safety problems and can distract students from thoughtful inquiry. When feasible, the opportunity to explore through microscale work should be provided. It encourages more careful observation and allows for students to "see" what is happening.

Assessment

Moving from experiences and words to symbols will be a difficult leap for students. We who design tests in chemistry should consider the ability to understand and use symbols as a separate variable from conceptual understandings. Some students will demonstrate achievement best by drawing or speaking; others will be able to use symbols. Good assessments will incorporate a variety of levels of questioning.

CD-ROM imagery and video experiments can provide prompts for questions about chemical phenomena. Most students will show greater achievement when they have images to help them access prior experience.

Personal and Social Perspectives

While studying historical discoveries, students can identify with the struggles of scientists to communicate new ideas in the face of resistance. This interplay between the ideas of scientists and the beliefs of society often personalizes many issues for adolescents and lays the foundation for recognizing that models emerge, are tested, and persist only until new technology forces them to be reevaluated. Models continue to be valid, are discarded, or become building blocks for new models.

Chemistry applications in the marketplace, our personal reliance on chemical products, and safety and disposal issues should be integrated into abstract lessons to make them relevant to the everyday lives of students.

Encouraging students to explore the economic and social parameters of industrial production and how industry conducts research will open students' eyes to the differences between industrialized and emerging economies. Inspiring a global perspective in tomorrow's citizens is an important pathway to scientific literacy.

Science and Technology

The history of chemistry offers clear examples of the relationship between the development of technology and the discoveries of science. Asking students to repeat the same experiment with a spring scale, a triple-beam balance, and then with increasingly sensitive electronic balances can convey a sense of the power of technology as a partner to science.

Using interfaces and computer probes to obtain information encourages students to focus on what is happening and to seek understandings about reactions among molecules. Working with data obtained quickly and in real time compels students to look for patterns. Students have more time to innovate, to explore alternative reactions, or to ask "what if" questions that extend their comprehension of the fundamental theories.

A Classroom in Action

Ray Williams is an "alumnus" of the Sputnik-era teacher training programs. For most of his career he has used a single candle flame as a starter for a lesson on observations. But now Ray has expanded many lessons on combustion, incorporating methods of inquiry, to illustrate the molecular nature of gases.

Students begin, as they have for decades, by observing flames. "How does the flame of a candle differ from that of a Bunsen burner?" The candle flame has three main reaction zones, a primary zone where combustion originates, a main reaction zone where combustion is completed, and a luminous zone where free carbon burns and luminesces. In the Bunsen burner the zone nearest the jet is a non-combustion zone, where air mixes with gas. The primary and main combustion zones still exist, but the combustion of natural gas is completed in the outer mantle where very little carbon is left to luminesce.

Students compare their observations to molecular models and to structural formulas representing the combustion of wax and natural gas. Because some of Ray's students still have difficulty relating physical observations to formulas, the class spends extra time conducting cooperative group discussions. Verbalizing misconceptions and questions helps all students gain content knowledge about the molecules that combusted.

Students compare flame color to temperature, using measurements, models, and diagrams. They form hypotheses about the relationship of color to the temperature of flame, and then test them by measuring the heat released by various types of candles and different aperture sizes in the Bunsen flame. They also vary the amount of oxygen in the Bunsen flame and record observations about color, heat, and amount of soot (carbon) produced by the flame. Finally, students are asked if there is a relationship between the amount of light a flame produces and the amount of heat released.

Most students can determine an inverse relationship, and the class moves quickly from observation to questions and investigations.

Like each of Ray's physical science lessons, the concepts learned in class are applied to everyday life. He presents the class with a camping lantern and asks them to determine the purpose of the mantle. It is usually relatively difficult for students to make the conceptual leap from textbook and simple experiment to this application; therefore, in most of Ray's classes, several mantles are used before students relate their knowledge of carbon incandescence (in the third zone of the candle) and the burned material of the lantern mantle.

continued next page

Because Ray's students have learned the basic safety precautions involved when using flames, he can feel confident in offering them the challenge of developing their own procedures for an open-ended laboratory. He may ask, "Does the temperature of the environment affect the color, rate of burn, or light released from a candle flame?" Student groups are quick to realize that they need identical candles, safety equipment (including goggles), and measuring equipment (thermometers or computer-interfaced thermistors, light meters) to do their work. They have progressed from observation through directed investigation to open-ended experiments many times before. They are able to develop procedures and conduct experiments with little direct supervision.

Ray moves through the classroom with carefully guided questions to ensure that students are not leaving any important consideration aside. His questions act as embedded assessments of student progress and help Ray pace his instruction, modifying it as necessary to match student understanding.

Quite often Ray's assessment for a chapter or concept is an authentic extension of the classroom experiences that occurred during the unit. For example, part of the unit test on flames asks students to determine whether a glass "chimney" makes a candle flame burn brighter or less bright, and then to use concepts and lessons from the chapter to explain the result. Almost all his students can safely and accurately determine that a chimney makes a candle produce better lighting, and the majority of students can relate the reduction in air currents around the candle to the increased luminosity. Only a few students can also infer the difference in oxygen flow around the base of the flame to the brighter candle. Because the assessment has many levels of "correct" answer, Ray can use it with students of varying abilities in the same testing environment.

While Ray's physical science classroom still relies most often on very simple phenomena, student achievement is high because each lesson follows the same learning cycle. Whether his students are burning candles, boiling water, dissolving salts, or examining common materials, Ray continually gives them a chance to understand everyday molecules with greater insight than before.

Resources for the Road

Flames Poster. (1991). Salt Lake City, UT: National Energy Foundation.

Lyons, John W. (1985). *Fire.* New York: Scientific American Books.

Structure and Properties of Matter

Nature of the Learner

Secondary students can make accurate observations of physical change; many will have difficulty relating these changes to molecular structure and the movement of electrons.

While most students will be able to describe the properties of matter, only a few students will be able to relate new observations to an understanding of structure.

Students' prior experiences may have resulted in misconceptions that will have to be carefully identified and confronted.

History and Nature of Science

Most textbooks show chemical bonds in distinct categories, but today's science has demonstrated that they are far more complex. New states of matter, such as the Bose-Einstein condensate, do not fall into the discrete categories of textbook chemistry. Sharing the latest research with beginning students can prevent them from becoming dogmatic about their understandings.

Because the nature of solutions, changes in states of matter, and bonding are concepts that can be studied at many levels, we should take care not to present the path toward understanding as a movement from "wrong" to "right." Oversimplification should be avoided. It is helpful to point out that new technologies allow scientists to make more finely tuned predictions about atomic structure.

Nature of Instruction

Rather than relying on single examples, we should encourage students to generalize from many experiences.

Building three-dimensional models will help students visualize molecular structures; manipulating these models will help them visualize molecular change. Such experiences allow students to comprehend the movement of electrons.

Laboratory explorations in the Standards-based classroom may be simpler than in the past, but they will take place much more often. Investigating 10 examples of changes in different solutions and comparing the visual and physical changes to find a common principle is more valuable for constructing knowledge than completing a single multi-step laboratory.

Comparison of results among cooperative groups offers a broader range of data to use to support critical analysis of what may be happening in a particular experiment. Cooperative groups also encourage students to communicate their understandings.

Assessment

Authentic assessment should follow instruction as closely as possible. Where students have studied several examples of a phenomenon, a new example makes a great assessment. We might, for example, ask students to identify unknowns based on their properties. Practical experiences don't always have to be associated with the test itself; many teachers are assigning labs as homework *before* the test. Students bring their individual data to the test on the appointed day and then use that data to make inferences.

We should encourage students to explain their understandings in words and drawings, in addition to answering traditional mathematical and analytical multiple-choice questions. Fashioning multi-leveled tasks, adopting a project-oriented approach, and encouraging many different styles of explanations allow for different student learning styles.

The most difficult level of assessment asks students to apply a classroom concept to a real-world situation. For example, if students have studied the freezing point of various salt solutions in class, the test might ask for an application: "Houghton, Michigan, is one of the coldest areas of the United States. But they don't use salt there to clear the roads of ice. Why?" This sort of problem-based assessment is especially appropriate for cooperative groups.

Personal and Social Perspectives

Students' homes and workplaces are filled with examples of physical and chemical changes. Providing students with home experiments can involve parents in the learning process. Investigations with home chemicals provide the most relevant examples.

Look for community-based problems so students will use parents and neighbors as resources. For example, what products or processes are involved in local industry? What effects do local industries have on the environment?

When nonstandard materials are used in the classroom, care should be taken to obtain data safely and to begin with a clear understanding of what the chemicals contain.

Science and Technology

Students can use many computer software packages to develop three-dimensional molecular models of what they believe exists as well as to test hypotheses. With practice, these experiences can move beyond exploration to become tools for increasing understanding.

Technology is also a valuable tool for students to use in gathering data. Particularly useful are hand-held probeware and graphing calculators.

The technology of industrial processes is rarely covered in secondary textbooks, but chemical engineering is one of the most promising career areas for today's students. Guest speakers and site visits can bring the technology of industrial processes to the classroom.

A Classroom in Action

Brian Murfin mentors teachers in New York. Many of his teachers need examples to help students appreciate historical methods; Brian suggests presenting information about traditional African iron-smelting techniques.

The earliest evidence of smelting metals comes from Saharan Africa (Nubia) around 4000 B.C. Iron was being smelted on the African continent long before it was discovered in Europe.

To encourage student interest, Brian often suggests beginning with pictures of early implements—spears, axes, and knives. Depending on the level of the class, student groups may be assigned to research drawings or descriptions of early civilizations or to read and report on descriptive articles he has distributed. Students are then regrouped and asked to research chemical questions about iron smelting:
- What is the reducing agent?
- What reactions are involved?

Students begin a laboratory activity that simulates iron smelting. They mix small amounts of iron (III) oxide and carbon in a crucible and heat the mixture for several minutes. They test the heated substance with a small magnet. Then they return to their groups to examine iron, brass, and steel tools on display. More questions are raised:
- Why did iron (rather than other metals) become so widely used?
- Did all cultures develop iron smelting?
- What uses can be made of the slag (byproduct) from smelting?
- Where are the largest iron deposits on the continent?
- How can iron be separated from iron (III) oxide?
- What would be needed to build an iron-smelting furnace?

Once students understand the physical requirements of a smelting furnace, they are often surprised to learn that early African metalsmiths often used the silica and alumina particles from termite mounds to line

continued next page

Resources for the Road

Childs, S., and Killick, D. (1993). Indigenous African Metallurgy: Nature and Culture. *Annual Review of Anthropology, 22,* 317–337.

Murfin, Brian. (1996). An African Chemistry Connection. *The Science Teacher, 63*(2), 36–39.

their furnaces.

Brian's colleagues have reported that these history lessons are especially appropriate for some learners, while other students prefer to move directly to laboratory investigations. Using cooperative group techniques, such lessons capitalize on the differences among students and multiply the interest of the class in the properties of matter.

Chemical Reactions

Nature of the Learner

Students may come to chemistry lessons with strong preconceptions about chemical reactions, believing they are always explosive, exothermic, and exciting. Instruction must be thorough, and each conceptual area must be explained as new experiences help students construct new knowledge.

Most secondary students will be able to understand that chemical reactions can dissociate molecules into constituent atoms, but they will have greater difficulty constructing accurate understanding about double-exchange reactions. Only a few secondary students have accurate, consistent models of conservation of matter and energy.

History and Nature of Science

Belief in alchemy is found not only in historical vignettes, but also in the minds of many adolescents who may have been exposed to caricatures of chemical reactions in the media. Those of us who use the history of science as a framework for modern discovery will find that many students resort to medieval explanations of chemical reactions.

Chemical reactions can take place in periods ranging from a few femtoseconds to millions of years. Expanding students' understanding of reaction rates and catalysts will require more laboratory time than in previous courses. A Standards-based classroom will put less emphasis on mathematical models that predict rates and more emphasis on developmentally appropriate discovery.

Nature of Instruction

Because the content understandings in chemistry are difficult for many students to achieve, instruction will proceed slowly in the Standards-based classroom. We will avoid "pop, bang, and fizz" experiments and demonstrations and use everyday events to illustrate concepts. We will also emphasize microchemistry and careful observations. Micro-quantities encourage students to analyze their own expectations more precisely and to verbalize interpretations rather than be led by expected results. Using the smallest possible quantity of reagents encourages both safety and more careful observation.

Standards-based laboratories will have fewer steps listed in "cookbook" fashion and will often suggest a general question and ask students to design investigations based on manipulating sets of variables and using available chemicals and equipment.

Assessment

Assessments should emphasize analysis of experimental results or the interpretation of chemical reactions in student-designed experiments.

Asking students to draw or model what they believe is happening in a chemical reaction can provide more valid forms of assessment than using mathematical formulas to make calculations. Few students at this level will be able to fully express their understandings in mathematical symbols.

We need to avoid tests that emphasize jargon and that ask students to solve stoichiometry problems from a pattern (or algorithm) that is common in a single chapter. The ability to follow patterns seldom indicates true understanding in chemistry.

Whenever a physical example or prop can be incorporated into an assessment, it will increase the validity of the test. A number of studies have established that even when an individual performance assessment is not possible, allowing students to examine the equipment or reagents that they used (or having them displayed at the front of the room) can help students demonstrate their understandings more effectively.

Personal and Social Perspectives

Chemical reactions can make headlines—ozone holes, fossil fuel shortages, smog, and toxic waste—and many fitness and health products that promise pseudo-medical claims. Consumer evaluation of products relates directly to an understanding of chemistry. Students who become critical thinkers and problem solvers can apply their ability to analyze chemical experiments to their own lives and to Earth's environmental problems.

Science and Technology

Once again, probeware can be used to collect highly accurate data that cannot be acquired by direct observation. Linking the data with computer analysis programs gives students opportunities to evaluate and manipulate data to understand chemical reactions.

While seldom available in classrooms, the newest technologies in detecting trace chemicals and the research that measures the rate of extremely fast chemical reactions can provide important insights to students.

We might also consider asking students to collect samples (water, pollutants, foods) and send them to a commercial or government testing lab to determine levels of chemicals with advanced sensing technology not available in the classroom.

A Classroom in Action

Judy Waldraff designs classroom strategies and adapts her traditional text material to be consistent with her constructivist approach. The work takes many hours after school, but her successes have given her peer teachers' different perspectives on what can be done. Judy shares her work with her colleagues, who in return help her set up some of the traditional lab equipment with which they are more familiar.

The differences between the traditional approach and Judy's can be seen in a classroom exercise involving the ordinary reaction between calcium chloride and sodium hydrogen carbonate. To initiate inquiry, Judy begins with a class brainstorming session. "What do you think will happen?" "How will you know it has occurred?"

Then Judy's students work in groups to discuss possible products and to propose an equation that will accurately predict the chemical reaction. As students present their ideas, Judy uses teachable moments to elucidate "testable" logic and ideas for good laboratory procedure. She reminds the students that they will not be allowed to conduct unsafe work in the lab.

Some students become frustrated; they want directions on exactly what to do, and they want to begin doing it right away. It is here that Judy has a chance to suggest that the process each group is experiencing parallels that of professional research scientists. For example, much thought, organization, and perseverance are required in chemical engineering before an experiment begins, and virtually all industrial research is done in groups.

continued next page

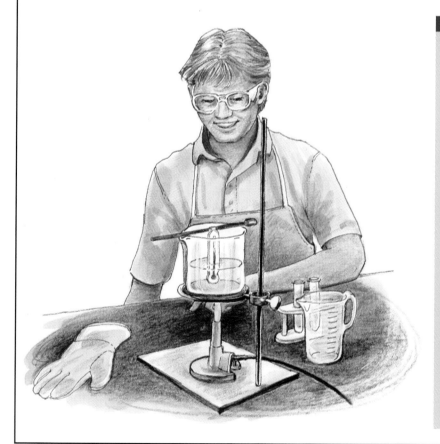

Resources for the Road

Clough, Michael P., and Clark, Robert. (1994, February). Cookbooks and Constructivism: A Better Approach to Laboratory Activities. *The Science Teacher,* 61 (2), 34–37.

Clough, Michael P., and Clark, Robert L. (1994, October). Creative Constructivism. *The Science Teacher,* 61 (7), 46–49.

Novak, Joseph. (1991, October). Clarify with Concept Maps. *The Science Teacher,* 58 (7), 45–49.

Streitberger, H. Eric. (1994, September). Modeling Molecules. *The Science Teacher,* 61 (6), 46–48.

> By having students explore possible chemical reactions, Judy seeks to develop cognition. Her methods include having students think out loud; develop alternative explanations; participate in cognitive conflict; access, retrieve, and use conceptual and stoichiometric information; set individual goals; make decisions; self-evaluate to solve problems effectively; design and test alternative hypotheses; and communicate their findings and understandings.
>
> Once Judy's students begin their experiments, they already have clearly defined expectations and hypotheses. Sometimes they are surprised by the result, but often they are gratified to have their planning and expectations confirmed. Careful planning allows students to develop confidence in their ability to understand science.
>
> Like Judy, we can modify our behavior and adapt existing activities to engage students in inquiry, open a window into their thinking, and advance along the pathways toward the Standards.

ematical information about the motions of objects in space accessed from CD-ROMs, multimedia databases, and astronomy simulations will contribute to students' knowledge about motion in the universe.

Our understandings about motion and forces have increased in direct proportion to the precision of our measuring instruments. Students can benefit from comparing the results of the same experiment measured with stopwatches and photo gates and with beam and electronic balances. They can extend their understanding by comparing images from optical telescopes and data from radio telescopes.

Motions and Forces

Nature of the Learner

This is the area of physical science in which students will have the most persistent preconceptions because every child experiments with motion and forces. Many naive ideas can be identified through concrete, student-guided activities.

At the secondary level, students can usually understand the concept of static forces and can measure the results of experiments reliably. Concepts of motion and conservation of energy will be difficult for most students. They will have persistent misconceptions about the need for an unbalanced force to keep bodies in motion, projectile motion, the interaction of forces in situations involving equilibrium, and the motion of satellites and objects in space.

History and Nature of Science

Because planetary motion is difficult to explore with hands-on experience, students can benefit from classic experiments, "How do we know?" questions, and from conceptual challenges ("Convince me."). Visual and math-

Nature of Instruction

In new physical science programs, more emphasis will be placed on electricity and magnetism as aspects of the electromagnetic force. Students also will explore more examples of the same law or phenomenon rather than focusing on a single concept in a discrete amount of time, then moving on to another concept.

Despite the fact that secondary students often seem quite logical and capable of abstract thinking, it is always appropriate to begin the study of a concept with a laboratory in which students gather data through the most direct means (visually, if possible) before moving on to computer-interfaced measuring devices. From these experiences, students can develop their own questions to investigate.

Assessment

Students often learn to express the "correct" answer in words while still maintaining misconceptions. To counter that tendency, we might ask students to draw the predicted path of a projectile and to identify what happens to forces and velocities while the projectile is in motion. Or they might extend the curve of a graph they have constructed and discuss the physical meaning of the slope. Or they might predict the result of a change in force. Such assessments are much better than asking students to repeat a definition or solve an equation.

We might consider creating real-world scenarios and asking students to apply what they have learned. For example, "How do you aim an arrow to hit a bull's-eye?" "If you downshift on ice, why does the car spin?" Pick a simple toy: "How does it work?" Whenever possible, allow students to watch a video or video clip or play with an object as they think. Also, computer-assisted demonstrations of student understanding can add a dynamic visual component to assessments.

Personal and Social Perspectives

Having students learn how they learn is an important part of scientific inquiry. Motions and Forces is a good content area in which to explain to students how they form ideas and how they can change their ideas. Historical examples of particular scientists who changed ideas for many and who built on the ideas of their predecessors can help students identify changes they may need to make to adjust their knowledge to be in line with current scientific theories. On the practical side, students might investigate how, for example, the laws of force and motion apply to their everyday lives.

Science and Technology

Emphasizing electricity and magnetism through applications with motors, generators, thermostats, and other common household objects offers safe, effective lessons on relatively simple technologies that are still mysteries to most adolescents. Effective lessons in electricity and magnetism also lead into asking questions about energy demands and the use and allocation of natural resources.

Connections with science, society, and technology are made as students are challenged to consider their personal energy needs and how this affects energy conservation. Alternative power sources can be explored. Microwave technology can be compared with traditional kitchen appliances to expand the realm of inquiry from the classroom to the home.

The electricity-and-magnetism connection also opens the window to explore transportation devices of the future, electromagnets, and magnetic levitation.

A Classroom in Action

Gina Falcone has been adapting student activities based on Workshop Physics and Real-Time Physics, two software programs that she studied at summer workshops at Dickinson University. It was serendipitous that Gina's school system wanted to integrate the use of technology into science classrooms at the same time that Gina wanted to institute a change in her teaching.

Since her school system is only an hour's drive from Dickinson, Gina has been able to get support from the college faculty. She also has been able to get grant monies to hold two-week workshops at her own school in order to share the Workshop Physics approach with teams of high school science and math teachers throughout her community. Opening a window into her classroom is one way that Gina shares her insights with her colleagues.

Gina had already introduced inquiry-based learning into her classroom. Constructivist-oriented materials and approaches have been interspersed throughout her curriculum during the past decade.

She still begins the year with the study of motion. The Workshop Physics and Real-Time Physics objectives and philosophy challenge teachers to design activities in which students have direct experience of motion and forces. Using new computer tools, Gina's students actively participate in quantitative explorations, asking "What if" questions as they master a traditional body of knowledge.

Using motion detectors, force probes, and self-propelled fan carts, they confront their personal conceptions about force and motion and compare them with experimental results. Using the new technology also allows students to make observations, gather data, and analyze results more rapidly and accurately. This gives Gina more time to interact with students to discover what they know. She tries to assess the depth of their comprehension through oral and written work. Strategies to assess student understanding vary.

Gina begins units with friendly concept pretests that are not graded to establish a baseline for students to predict what they think will happen in given situations. Students revisit these predictions as they conduct a series of investigations.

They design models to explain what is happening in particular situations. They cooperate with one another and share ideas about phenomena. As they look at results on the computer and try to figure out what they mean, Gina allows for wait time. While the students discuss their findings, Gina asks leading questions to direct their thinking. She encourages her students to consider all possibilities, then examine them for flaws.

As they examine their data, students begin to synthesize information. Having

continued next page

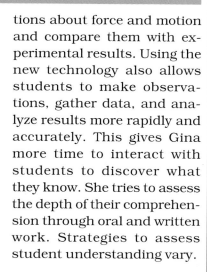

gathered important statistical information, students are able to communicate their understanding through the presentation of mathematical models.

Computer-interfaced data collection takes the tedium out of many labs, encouraging more speculation and risk taking in analyzing the results.

Finding time to reflect on the changes in her classrooms is a priority for Gina. As she modified her curriculum, she had to alter her assessments so she could determine whether her broader goals were being achieved. Now her students demonstrate their achievements by creating portfolios and designing experiments and projects, in addition to taking traditional tests.

On a path toward implementing the National Science Education Standards, Gina stands in the forefront because of her willingness to adapt new curricula, technology, and assessment packages.

> **Resources for the Road**
>
> **Real-Time Physics** (software by David Sokoloff, Ron Thornton, and Patricia Laws). Portland, OR: Vernier Software.
>
> **Workshop Physics** (software by Priscilla Laws). New York: John Wiley.

Conservation of Energy and Increase in Disorder

Nature of the Learner

As formal reasoners, most adolescents should be able to understand the conservation of energy in its simplest form. They will have more difficulty understanding the concept when energy is transformed into multiple forms and when it is related to energy transfer with an increase in disorder.

Because statistics is rarely covered in many secondary math courses, the concepts of random motion and disorder will be difficult for most high school students. It will take significant practice and experience for students to construct lasting knowledge. We may need to teach some topics in simple statistics and probability.

History and Nature of Science

The transfer of energy has its most relevant application in thermodynamics and mechanical engineering. These topics can easily be introduced at a qualitative level, but will be difficult to explore at an accurate, quantitative level. Today's research in energetics relies on advanced mathematics and chaos theory, as well as mathematical modeling at a level that will challenge even the best students.

Nature of Instruction

Energetics and entropy are among the areas in which the level of mathematics necessary to do research in science is so far beyond the level of understanding of students that secondary programs often resort to analogies, interpretations, and examples that more or less imitate reality. Our challenge is to create relevant explorations that are true to theory while offering understandable applications, and to avoid mythologizing the advanced mathematics in modern physical chemistry.

We should carefully avoid assertions and direct instruction. Conservation is a concept that must be discovered to be truly understood by adolescents. Many simple experiments can be presented from which students can generalize. Cooling coffee with cream, heating an aquarium with a thermometer, and putting a cold drink in a cooler are all situations in which students can measure the conservation of energy without elaborate equipment.

Assessment

We need to assess carefully students' ability to comprehend conservation, beginning with simple examples. Conservation of energy is a logical scheme that should be available to most secondary students, but those who have had limited opportunities to explore may have very concrete ideas and persistent misconceptions. Secondary students should be able to analyze data tables and provide written or oral descriptions of the cumulative results. Test items can be written that involve practical examples, including windows, doors, and kitchen appliances.

Personal and Social Perspectives

Energy conservation is a key economic and social issue that students can explore. Mechanical engineering provides an extension to career education—students seldom get opportunities to experience what a career in engineering would involve. Engineering requires teamwork.

To mimic the work of engineers, we might consider assigning group problems that involve energy conservation in the school environment:
- How much heat is lost when the front doors of the school are opened?
- Could solar panels on the roof of the school power our computer lab?

Science and Technology

Thermodynamic experiments are best accomplished by computer-assisted measuring devices. Gathering large quantities of data over long periods of time can illustrate the value of today's instrumentation to students whose previous experience may have been only with thermometers.

The technology for the group engineering experiments suggested earlier can be easily constructed by students or incorporated into calculator-based laboratory (CBL) equipment. While some schools are discarding old Apple computers, physics classrooms still find them great for practicing interfacing. Measuring devices interfaced to calculators can make technology affordable.

A Classroom in Action

The Standards urge us to teach the personal and social perspectives of science—and there's nothing more personal than the family budget! In many classrooms the abstract concepts related to energy conservation become practical lessons about saving money at home.

John Lee's inquiry-based approach to energy conservation usually begins with experiments using model laboratory systems such as calorimeters. Exothermic chemical reactions can be controlled and measured, and students can follow the transfer of energy from one form to another.

Like other instructors, John follows such abstract investigations by asking students to build models of rooms or homes in which they can use various materials to control the transfer of energy from light to heat and the conduction of heat energy. Scientific modeling is usually a mathematical, abstract process, but physical models are particularly appropriate in engineering. Advanced students often make the conversion from physical models to computer spreadsheet simulations with ease, and experience engineering first-hand.

Now for John's students, along with more than 300,000 other secondary students, the practical implications of energy conservation extend a step farther, as they take part in a program called In Concert with the Environment.

Developed by EcoGroup, Inc., this program is usually sponsored locally by the area

continued next page

utility company. Participating students begin by conducting home energy surveys to examine the physical components of their homes—insulation, appliances, construction, heating and cooling equipment, and water faucets.

The program also includes investigations into physical science concepts, such as energy conduction by various materials. A sample laboratory examines the insulating capacity of single versus thermo- (multi-) paned glass. Students measure energy transformations (light to heat energy) and energy transfers, and they compare this data with the results of their home surveys.

Students translate real-world data into a computer model through a program provided by the In Concert with the Environment curriculum. The result of their work is an action plan, which they can share with their parents.

Moving physical science concepts into relevant, personal action plans takes a classroom a long way toward the Standards. Programs like In Concert with the Environment also illustrate how we can integrate mathematics and social sciences into science and put students in touch with the methods of mechanical engineering.

To get a list of utility companies that sponsor the In Concert with the Environment program, write Eco-Group, 5030 E. Sunrise Dr., Phoenix AZ 85044-5201.

Resource for the Road

Vernier, David L. (1986). How To Build a Better Mousetrap. Portland, OR: Vernier Software.

Interactions of Energy and Matter

Nature of the Learner

At the secondary level, most students can reliably gather quantitative data about wave action and make predictions based on that data. Students will need extensive practice to extend their understandings about one wave phenomenon to another.

Students often confuse nuclear and chemical energy. Once they study the loss of matter in nuclear reactions, unless their prior constructs are firmly in place, they may revert to their preconceptions that matter can be lost in chemical reactions.

History and Nature of Science

Physicists use the same mental models to explain visible and invisible electromagnetic radiation, physical (sound and seismic) waves, and nuclear wave phenomena. The mathematical principles connecting these functions will be beyond the scope of most secondary students, but they can see similarities among them.

Physical chemistry is making exciting progress in exploring new states of matter and new evidence for unified theory. We can discuss this progress with secondary students when appropriate learning foundations have been built.

Students will have little understanding of the scientific methods that teams of nuclear scientists (often linked internationally) use to design and conduct experiments. Films, journals, and Internet links can help students comprehend this modern scientific method.

Nature of Instruction

Sound and light have been traditional subjects of extensive exploration by secondary students; these topics are still central in the Standards-oriented classroom. Tomor-row's classrooms may put more emphasis on exploring radio and television waves and investigating (through library or communications research) the very long and very short wave phenomena that give scientists information about activity outside our solar system.

Assessment

The constructivist classroom is supported by good questioning. Assessment should occur throughout a lesson, not only at the end.

When an area of physical science involves so many phenomena that are invisible to students' senses or direct experience, evaluating student progress by constant open questioning becomes important. We need to ask good questions often and vary the level of the questions regularly. We should ask not only what students know, but also why they believe it and where they can apply it.

Personal and Social Perspectives

Exploring the use of radio telescopes to map the universe is an exciting bridge from concrete lessons to the imagination for most adolescents. Many students have found the route to physical science through the door of the imagination. Making difficult physics personally relevant by injecting the excitement of where today's lessons might take the next generation can be very appealing.

On the flip side is the pseudoscience of electromagnetic effects. Depending upon class interest, confronting superstition can also be a route to student interest. "Do high-intensity power lines damage human health?" "Do humans give off 'wave phenomena'?" Some teachers use the tabloids as starters for great classes!

Science and Technology

The technology of today's radio telescopes (like Hubble) is as close as the classroom modem. Accessing satellite data is an important and increasingly commonplace experience for students who take television satellite transmissions for granted without any consideration of the technology that supports them. We can ask students to compare early photos from the Hubble telescope to later ones from the repaired telescope. Then we might ask, "What could Galileo see?" "Kepler?"

A Classroom in Action

Joyce Marshall and John Rourke's classrooms reflect ideas promoted by the National Science Education Standards. Both teachers are taking advantage of thematic approaches to physics in which students explore real-life topics that are part of Active Physics, a program designed for high school students who have limited prerequisite mathematics and reading skills.

Students have opportunities to acquire depth of understanding in areas pertinent to the design and execution of challenges presented in each unit. As content knowledge is needed, students are offered a variety of ways to seek understanding in specific areas.

In the Home unit, students are challenged to design a prototype home that can be built in developing countries throughout the world by asking "What Do You Think?" questions such as

- "Does the shape of a universal dwelling influence how it gains and loses heat, and if so, how?"
- "What is the difference between heat and temperature?"
- "Are some materials better at storing heat than others?"
- "If you look at a car radiator or a home radiator, you will notice metal fins are a big part of the structure. What is the purpose of these fins? Why don't we just have a single piece of metal to serve the same purposes?"

These questions set inquiry scenarios that are sequenced with "For You To Do" investigations, "For You To Read" inserts, "Physics To Go" out-of-class assignments, and "Stretching Exercises."

Students learn about energy, electricity, and scale as they undertake the design of the universal dwelling. As they apply their learning to the design, they also must become decisionmakers about the effects of climate, availability of resources, perceived human needs, and other dimensions of science and human endeavors.

Assessment rubrics differ for each unit. Rather than written tests and problem sessions, or even traditional lab reports, the criteria for assessment emerge from the nature of the challenge. For example, in one section of the Communications unit, students are challenged by a scenario that begins, "Although virtual reality promises to make our sense of touch an integral part of our entertainment media, most of our entertainment today still comes from the communication of sound and light signals...."

The evaluation criteria for this task are based on creativity of design and explanation of the physics principles used to design, produce, and present a sound and light show to entertain classmates. The sought-after sound and light effects must be created with conventional equipment: lasers, simple circuitry to convert other forms of energy into sound, color filters, diffraction grating, Polaroid film, the human voice, and various materials to build instruments. The scope and depth of student understanding of the wave properties of sound, light transmission and interference, and some electronic circuitry form the foundation for students to apply their learning to an active situation.

Active Physics, an alternative physics course for grades 9–12, is an NSF-supported curriculum project developed by the American Association of Physics Teachers with assistance from the American Physical Society.

Resource for the Road

Active Physics. (1996). College Park, MD: American Association of Physics Teachers (AAPT)

A Classroom in Action

Integrating the Sciences

Current health topics, including relationships between salt and hypertension, cholesterol and heart disease, and fiber and cancer, are part of a unit on the Chemistry of Fitness. Working with two other teachers, David Rodriguez modified an existing three-week physical science module to include an introduction to organic chemistry, the biochemistry of nutrients, food chemistry experiments, and some basic information about exercise. This allowed students to qualitatively and quantitatively explore issues of health, fitness, and diet while they learned chemistry.

Students enthusiastically say that this unit, similar to ones in ChemCom, is interesting because it deals with tangible concepts relevant to athletic performance and body fat. Through inquiry, students gain an understanding of chemistry that supports personal choices (Bergandine, 1991).

Making science applicable to athletic endeavors or biomechanics is a way to capture student interest while promoting inquiry. Mary Ann Jenkins encourages students to think critically and to design effective experiments. Before studying motion by measuring time and distance for student sprinters, the class discusses various parameters and assesses their value to the scientific endeavor.

Students consider the process of gathering information about a sprinter's performance by asking, "How far apart are the timing gates?" "What quantities need to be measured in order to calculate the sprinter's velocity and acceleration?" "What would be the best method for obtaining accurate data?" After the data is collected outside on the school track, the students use computer spreadsheets, or perhaps graphing calculators, to analyze sprint motion.

Mary Ann prompts the students to consider "boundary" questions, such as "Which runner had the greatest/least average acceleration through the first gate?" "Which runner had the greatest/least average velocity through the whole course?" "Which runner had the smoothest/most irregular acceleration through the whole course?" Mary Ann further prompts the students to identify, describe, and evaluate sources of error (Slaughter, 1991). Students gather data using stopwatches and meter sticks, or the student athletes strap one of the newer CBL units with an attached accelerometer to their backs and actually "measure" acceleration.

Darrell Wilson expands the study of motion and associated forces to many athletic activities. With a video camera, a VCR with slow-motion capabilities, and digitiz-

continued next page

Resources for the Road

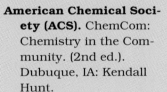

American Chemical Society (ACS). ChemCom: Chemistry in the Community. (2nd ed.). Dubuque, IA: Kendall Hunt.

Bergandine, David R., Jones, Loretta, and Federick, Betz. (1991, March). The Chemistry of Fitness. *The Science Teacher, 58* (3), 29–32.

Haddad, Ron. (1993, February). Biomechanics and Biceps. *The Science Teacher, 60* (2), 23–25.

Harley, Jr., David W. (1990, September). Track and Field Work. *The Science Teacher, 57* (6), 58–60.

Slaughter, Michael G. (1991, May). Sprinting into Science. *The Science Teacher, 58* (5), 41–43.

ing software, Darrell provides students the opportunity to eliminate some sources of error by making more accurate measurements on track and field events or on basketball and volleyball courts.

After completing an in-depth conceptual and mathematical analysis, students share ideas with the class through explanations supported by diagrams and visual aids. During these presentations, students examine variations or propose hypothetical situations to determine optimum efficiency and performance. The challenging tasks of taking science to the practice field or court furthers the understanding of physics. Perhaps athletic performance is enhanced through these teaching strategies, too (Harley, 1990).

Renata Holmes takes a step toward the Standards by *bringing the physics of the human body into 9th- and 10th-grade biology classes. Starting with a brief reference to mechanical advantage and the principles of levers, including simple torques, Renata asks students to solve problems in biomechanics, specifically ones associated with the biceps. Working individually with private questions directed only to the teacher, students are challenged to calculate the minimum force the biceps must exert to move a certain object upward.*

How does a teacher offer guidance to all, especially when some students quickly devise an approach to the challenge while others may be puzzled, frustrated, and say that they are clueless? Renata emphasizes to the students that the process of seeking a solution is far more useful than an explicit numerical calculation arrived at by using a skeleton and measuring instruments. Patience, perseverance, thoughtfulness, and imagination should prevail. Once students meet the challenge, some with Renata's guided intervention, she and her students evaluate the proposed solutions and determinations.

This problem in biomechanics could be expanded to encompass vector analysis, forces associated with the forearm positioned at varying degrees of flexion, and graphs of variables (Haddad, 1993).

Imaginative teachers looking for ways to integrate material from the Standards can emulate these types of teaching strategies to adapt studies of skeletal systems, the physics of sports and dance, and chemistry and nutrition to reinforce science content and develop a habit of inquiry in their students.

Life Science

For decades, many of the nation's life science classrooms have been anything but lively. Biology has been criticized for being content-heavy, overloaded with vocabulary, and tested by rote. Six to seven hundred pages of text, presented to teenagers with limited abilities to reason, constituted what in most cases was the only required science in high schools. By contrast, the classroom of a life science teacher who is moving toward the Standards provides an exciting environment for inquiry and a core of content that is smaller but in greater depth than in the past. Topics are covered in more detail, coursework is integrated, and both teachers and students feel challenged.

New content is the primary force driving a Standards-based life science curriculum. Life science itself is changing rapidly. The Standards define six key areas that should be part of every secondary life science program:
- *The Cell.* Cells are the unit of structure and function of living things.
- *Molecular Basis of Heredity.* Organisms ensure continuity through the genetic code.
- *Biological Evolution.* Change through time has ensured adaptation to changing environments.
- *Interdependence of Organisms.* Energy and nutrients cycle through the ecosystem.
- *Matter, Energy, and Organization in Living Systems.* Producers store the Sun's energy in organic molecules, and consumers use that energy for life processes.
- *Behavior of Organisms.* Living things respond to stimuli in ways that are both genetic and learned.

For secondary students, the first content area, the study of the cell, is traditionally the first topic presented. Our challenge is to make connections between what students can see with the naked eye (and what concrete thinkers can understand) and what students can see under a microscope. One example is the relationship between structure and function. Just using the microscope can become a source of misconceptions for students who cannot perceive the reality of the view within the ocular.

In a Standards-based classroom, microscope labs will never stand alone. Macroscopic investigations will be paired with microscopic observations to help students bridge the gap from concrete to formal logic. At the same time we will not allow students to rely on rote learning (such as memorizing the names of organelles) to achieve success; vocabulary will be limited, but concepts will be expanded and extended. The cell will be studied as both a unit of life and a model for the structure/function theme, which is echoed in all life science content areas.

A key understanding in life science is the molecular basis of heredity. The relationships between genetic material, health, disease, and behavior are major issues not only in scientific but also societal contexts. Study of the DNA molecule provides a context for students to understand the nature of modern life science.

But this area of science is moving so quickly that the majority of us lack appropriate content training. In addition, the isolation of our classrooms, the limited nature of school lab facilities, and the difficulty of devoting time to postgraduate training present us with significant challenges in achieving the content standards in molecular biology.

To meet the Standards relating to evolution and the interdependence of organisms, students will need a solid background in mathematics. But this will be difficult to achieve given that biology, which has traditionally been the first secondary science, has not been perceived as a quantitative science. Even students who do well in math will have difficulty integrating those skills into modern understandings of evolution and ecology.

Similarly, studies of matter, energy, and organization of living things will require students to build on prior understandings of matter and energy. Consequently, today's biology can rarely be accomplished by ninth graders who do not have a strong physical science background. Therefore, to move toward the Standards, many schools are examining the sequence of courses in their secondary programs. Many innovative integrated science curricula, for example, pair exercises in inorganic chemistry and photosynthesis in the same year.

Including behavior as a key content area in life science will be new to many of us. In many classrooms, only one "scientific method" has been taught in the past. Now behavior studies expand our understanding of the nature of science and also often include studies of human behavior. Personal perspectives and the relationship between behavior and individual and community health will be new content areas in many programs.

The shift in life science content may be difficult for many of us for at least four reasons:
- Many of us have not been trained in the new content areas and will need the support of professional development.
- Students may have difficulty mastering the new content because of their level of development or their limited experience in mathematics and physical science.
- Many classrooms have been built and equipped for exercises (like dissection) that will be less common. They lack the facilities (such as gas, water, safe storage) to support the new inquiry experiences.
- We will have to forgo content areas in which our students have been "successful" through memorization and replace them with content areas that require higher levels of logic, mathematics, or other skills.

For traditional life science teachers, what may be most disturbing is what has been left out. The Standards include almost no taxonomy. Systematics is only covered to illustrate the interrelationship between molecular biology and evolution. Similarly, anatomy is only an illustrative tool for the structure/function connection. In the most common textbooks of the 1960s and 1970s, the majority of the content focused on biological nomenclature and descriptive taxonomy, and much of the rest covered anatomy. The reason was pragmatic: Fifteen-year-olds, who are largely concrete thinkers, could pass a vocabulary test. Only a relatively few students would move on to the more highly logical content of chemistry.

The Standards prescribe inquiry not only as a subject of study (the "way science is done") but also as a method through which content should be learned. This is especially important in high school biology because of the age and experience of most learners. The conceptual journey from concrete to abstract will be repeated for each of the six core content areas for the majority of 9th- and 10th-grade students. Many will experience formal reasoning, methods and patterns of logic, decisionmaking, and applications for the first time in high school biology. To empower them, we will have to give them the gifts of confidence, excitement, and most important, sufficient time.

Come through the door of tomorrow's life science classroom as it moves toward the Standards. Things look very different:
- Many of the students are 11th and 12th graders and more skilled, because programs will place biology after physical science or integrated with it. Classes are smaller to allow laboratory experiences for all.
- Classrooms are equipped with science tools, materials, and technological resources, including computers and access to networks.
- Biology students are using electronic measuring devices and statistical tools. Integration with high school mathematics is the norm. The emphasis on statistics in the National Council of Teachers of Mathematics (NCTM) standards has proven a positive one for life science programs.
- Facilities are available for safely maintaining living cultures. Observation of normal taxes (responses) and behaviors is an essential part of the curriculum.
- The physical arrangement of the classroom is more flexible. Old-style desks bolted to the floor and facing front are gone. The classroom emphasizes cooperative learning and discovery. The action is not directed toward a podium or focal point, just as the curriculum is no longer directed toward an unalterable set of facts.
- Flexibility extends to time. Block scheduling and cross-curricular connections are the norm. Isolated 45-minute periods aren't flexible enough for the life science classroom of the 1990s and beyond.
- Life science programs are clearly integrated with physical science and mathematics; in many curricula, the courses do not even have separate titles. Unifying Concepts (such as Systems, Order, and Organization; Evolution and Equilibrium; or Form and Function) are the organizers of multidisciplinary studies.

In the pages that follow, we will explore each of the six core content standards in detail, with a corresponding example of best practice. The vignettes were selected from thousands of exemplars in American schools.

A good way to begin to move toward the Standards is to reach out to a colleague. A convenient source of communication for biology teachers is the Access Excellence forum, supported by Genentech, on Internet at http://www.gene.com:80/ae/. Another is TERC at http://hub.terc.edu:70/1/hub/owner/TERC/projects. Our professional colleagues, wherever we meet them, make great companions on the journey to the Standards.

The Cell

Nature of the Learner

Adolescents who have had the opportunity to explore in the context of the life sciences should be competent at making careful observations through the microscope and making observations and measurements of macroscopic organisms.

As 9th and 10th graders explore life science, they will continue to improve their ability to relate microscopic observations to hypotheses and inferences and to design experiments and analyze data. They will begin to understand models and simulations, make predictions based on the results of experiments, and relate chemical and molecular structures to observations about structure and behavior.

At least half of the students in beginning biology classes will have occasional difficulty with formal reasoning in this area. For example, they may not be able to relate a text explanation of plasmolysis to a prediction about the effect salinity will have on a cell they can see.

History and Nature of Science

Historically, the study of cell biology has progressed from observation and classification to modern biochemical theory. Ninth and 10th graders will feel most comfortable with historical science (Hooke, Schleiden, and Schwann) and will need carefully guided experience to understand modern science. Once students master an idea by rote ("All organisms are made up of cells"), they often resist developing a broader understanding of the concept later in the year. This problem can be minimized by emphasizing the relationship between the limits of technology in each period of scientific discovery. "What was the power of Leeuwenhoek's microscope?" "What could he see?" "What would he have done differently if he had a modern compound microscope?"

Nature of Instruction

Biology classrooms moving toward the Standards place less emphasis on terminology and memorization and more on observation, inference, and synthesis. Fewer topics and more in-depth coverage are essential. This means we will put
- less emphasis on the anatomy of cells and organisms and more emphasis on the relationship between structure and function
- less emphasis on a model cell with every possible organelle and more emphasis on the differentiation of cell structures to accommodate function
- less emphasis on observational labs and more emphasis on investigatory labs

Assessment

Recognizing that students' capacity to reason on paper may be limited, we need to expand assessment to include more practical laboratory challenges and open-ended responses. Multiple-choice tests do not easily measure the outcomes of a Standards-based life science classroom. However, many hybrid testing situations, which involve simulations, labs, or video presentations followed by written response, have proven to be valid.

Because modern students have so much experience with simulated video images, they often have difficulty believing that what they see through the microscope is "real"—or simply tell teachers that they see what they believe they are supposed to see (from text images). This can be avoided by encouraging students to manipulate the cells they see under the microscope, and then discuss and draw the results of their stimuli (for example, stomata opening and closing in response to humidity; contractile vacuoles reacting to salinity).

Personal and Social Perspectives

Cell biology offers many opportunities for exploring the personal and societal implications of content. Some of these include
- variations in cell structure within populations
- predictions about the likelihood of disease from observations of cells and tissues
- implications of human actions on cell structure (for example, the effects of smoking on cheek cells)

Science and Technology

Our understanding of life science has increased in direct proportion to the growth of technology. Students should be able to understand the gains in understanding that accompanied the development of microscopes, ultracentrifuge techniques, radioisotope studies, and culture technologies. Unfortunately, the cost of most technologies is still beyond the reach of many classrooms. To circumvent this barrier, we might consider trying to establish partnerships with an industry or university.

A Classroom in Action

Every year for a decade, John Alvarez has begun his biology class with a lab on microscope work. The textbook has examples of such activities with cheek cells, micromeasurement, and *Anacharis* lysis. John's students complete the labs accurately most of the time.

But each November, when he brings out the microscopes again, he realizes that his students have difficulty relating the structures they see to functions. Because he believes that all students should be able to master cell biology objectives, John spends some time talking with his students and finds that their difficulties fall into three categories:

- *Students with physical limitations.* Some 15-year-olds can't make the fine adjustments necessary to focus a compound microscope on high power; others with strong corrective lenses can't focus through a monocular microscope. Still others can't coordinate sight and fine motor skills.
- *Students who are still concrete reasoners.* Many of John's students have very little difficulty filling out worksheets and identifying the parts of a cell from models. But many others have trouble relating what they see in the microscope to paper-and-pencil diagrams.
- *Students who are only comfortable at the knowledge level.* When John shows students cells that have undergone structural changes (lysed plant cells, endoplasmic reticulum increases caused by low-level toxins), these students have difficulty relating the change in structure to the stimuli.

Using the Science as Inquiry standards as a guide, John redesigns his biology laboratories to provide experiences that will help students form models by using logic and evidence. Reversing the normal sequence of instruction, he asks students each week during the fall to use evidence to build a model in some context. A diagram on the bulletin board reinforces the model-building process.

For example, in the unit about cell organelles, students begin by examining *Anacharis* cells under the microscope, drawing their observations as carefully as they can. Then they add 5 percent saline solution to the cells and again draw what they see. They work in groups to respond to these questions: "How many layers are on the outside edge of each cell?" "Which of the layers is stiff?" "Which is affected by salinity?"

The students then draw or build models of what they see. They research their models by making more observations, viewing slides of other types of cells, and reading.

For students who have trouble seeing through the microscope, John uses a variety of technologies, including a small video camera hook-up (a simple security camera mounted vertically) that students can engage or remove to compare their own observations to those of others and a videodisc of slides they can refer to when discussing their observations.

continued next page

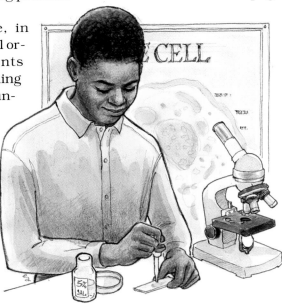

Life Science

For each cell biology concept in the fall, John's students follow the same path of inquiry. They are always reminded to explore first. Biological terms are introduced much later in each unit. John uses drawings and interviews throughout to search for misconceptions.

John's principal has supported his efforts to move toward the Standards even though John told him that he would only cover about two-thirds of the content he had previously covered in the fall semester. Together they examined the textbook to determine which chapters should be omitted to allow this slower pace. Then they informed parents and the school board about the change and included a summary of the Content Standards.

When John gave his students their midterm exam in January, he discovered that in relation to students in previous years,
- This year's students retained more knowledge and had a better understanding of concepts from the fall.
- They were better able to relate structure to function.
- They were less confused by terms that were semantically similar (for example, cell wall versus cell membrane).

Molecular Basis of Heredity

Nature of the Learner

Beginning biologists in grades 9 or 10 can usually follow directions in carrying out chemical or genetics experiments and can build models and make observations about their structure.

With practice, students of this age will improve their ability to draw inferences based on chemical reactions, isolate variables in experiments, and use combinatorial reasoning in genetics.

High school students will only be beginning to make predictions based upon probability, identify the information needed to solve genetics problems without prior models, and understand genetic control systems. Even when they can reliably solve problems in permutation and combination, probability, and statistics in mathematics class, they will have difficulty transferring that information to situations in life science. (We may have to help them make this transfer or, if they are ninth graders, teach them simple statistics and probability.)

History and Nature of Science

Today, genetics is based on chemistry, and students need to be able to relate the geometry of models to the behavior of unseen molecules. Too great an emphasis on 19th-century genetics can prevent students from appreciating the implications of modern genetics.

For example, once students are able to predict the ratios of Mendelian traits, they often overgeneralize and believe that many or most traits in humans can be predicted this way. This overconfidence can become a barrier to further understanding. The work of historical figures should be used to illustrate processes (such as statistics or geometric modeling) rather than as sources for facts.

Nature of Instruction

A Standards-based biology course needs to follow student experiences in physical science and must have mathematics integrated throughout of the course. Geometry and statistics can be reviewed and illustrated by genetics experiments. The methods of mathematical and geometric modeling will be new to most students.

While genetic engineering laboratories are commonplace in many biology classrooms, traditional work with *Drosophila* is still useful when teaching the value of mathematics, models, predictions, and inferences. Making a prediction about three generations of organisms and seeing it verified (within predetermined confidence intervals) successfully integrates mathematics and life science.

Assessment

One of the best uses for historic data in genetics is in assessment. Examples might include Mendel's data, Chargoff's experiments, or plots of fossil statistics. Students should be able to find patterns in data and to understand the range of variation due to chance. We can either provide pre-selected samples of offspring (for example, bags of peas in approximately 9:3:3:1 ratios) for students to analyze and explain, or we can provide data tables.

A less appropriate method of assessment is a human pedigree; very few human phenotype characteristics are truly Mendelian. Asking for definite interpretations of human characteristics can lead to overgeneralizations.

For students who have difficulty with formal reasoning, physical models can be important tools for assessment. Most of us are comfortable with laboratories where students use models, paper clips, or paper units to simulate DNA. That same material can be converted to an authentic assessment by creating model "genes" and asking students to decode them.

Personal and Social Perspectives

The study of genetics provides a basis for understanding the human condition, including developmental differences and disease. This can be vital

to adolescents as they explore their relationship to others and their personal best.

Many textbooks oversimplify the nature of gene action to provide situations in which students can solve mathematical problems. While the development of math skills may be important, it is more important that students understand variability and probability in relationship to genetics. The tendency of young biologists to look for absolutes will hinder understanding. A role-play of genetic counseling could be an application or an assessment. By asking students to assume the role of health care professional for an imaginary client, we can help them understand the difference between probability and inevitability and the importance of behaviors in determining health.

Science and Technology

DNA science is an empirical base for forensic science, genetic engineering, and discoveries in evolutionary science. Today's genetic technology has such widespread application that understanding the processes should be part of every secondary biology course.

The equipment required for students to perform genetic transformations or DNA fingerprinting is expensive. However, many teachers and schools have found great success by creating county consortia and purchasing laboratory equipment for advanced labs in common. One California county circulates all the equipment for genetic transformation laboratories among its schools all year, reducing the cost per classroom to a manageable figure. County school service agencies can be invaluable in arranging such shared-time equipment.

A Classroom in Action

Tarzza Jones's fellow teachers use videos and movies frequently. But the passive nature of watching movies and the time it takes disturb him. He wants to move his class toward inquiry, but doesn't want to abandon the information in the school video collection entirely.

Tarzza knows that at times his sophomores are concrete reasoners and they understand only vaguely the process of geometric modeling in biochemistry. Using materials from a variety of sources, he develops a unit in which students "rediscover" DNA, and he interweaves it with the classic BBC movie *The Race for the Double Helix*. His goal is to help students follow the logic and methods of Watson and Crick.

Day 1: Students extract DNA from fresh materials. They are encouraged to explore the viscosity and appearance of the spooled DNA. They describe their physical observations in a journal and use their imaginations to create "if-then" statements that relate observable physical properties to possible molecular shapes.

Day 2: Students take a look at Watson and Crick and explore what was known about DNA in their time. They also examine the tables from Chargoff's classic experiments. They develop a series of arguments to "prove" that genetic material is a protein and outline the physical characteristics of DNA known by chemists in 1950. The class views the first 15 minutes of the film, in which Watson is presented as a student looking for a research idea.

Day 3: An enlargement of Rosalyn Franklin's image of DNA is on the board. Class begins with speculation: "What can you see?" They watch the next 20 minutes of the film. Rosalyn Franklin's work is featured as well as arguments of biologists about the structure of DNA. Tarzza encourages students to identify with the scientists in the film, including Franklin. "Was it fair to use her work?" he asks. "Was she given appropriate credit?" For homework, students take home copies of Chargoff's data with ox and human thymus, spleen, and liver as well as Franklin's photos. Students are challenged to draw up a series of conclusions about what the data says about the molecule.

Day 4: It's time for model building. Tarzza asks, "Can you create a model that is

continued next page

geometrically compatible with the photo and statistically compatible with Chargoff's data?" "Can you design experiments that would test the accuracy of that model?" Students are given tag-board models of bases (from BSCS, 1989) and asked to construct a DNA molecule. (Procedures and limitations are presented in the BSCS text.) Students are challenged to develop additional "if-then" statements and to discuss how scientists would test this model.

Day 5: Students watch the end of the movie, and assessment begins. They take home one of two pages with sample research data from scientists who followed Watson and Crick:
- autoradiography and a one-sentence summary of results from Maurice Wilkins's 1956 paper (with photos similar to those of Franklin but with varying environmental conditions)
- photos of root tips treated with tritiated thymidine taken by Herbert Taylor (DNA picks up thymidine in the first generation; radioactivity is reduced by one-half in the second generation.)

Students are to answer two questions for discussion during the next class: "What do the results imply?" "Do the results support the geometric model or refute it?" Because members of each group have different assignments, they return Monday with much to discuss.

How is Tarzza's historical simulation different from watching a video about a famous scientist? Tarzza has identified which processes of the historical scientists are meaningful in today's science. For example, Watson and Crick's research illustrates the use of geometric modeling. Moving from photos to geometric models to testable inferences, and then inviting the scientific community to participate in the testing is a sequence that still works today. Moreover, many 10th-grade biology students will have concurrent experience with geometry; so, learning about the methods of Watson and Crick can reinforce the significance of many abstract geometric principles.

Tarzza has not ignored the maturity level of his students, who enjoy seeing scientists as interesting, valuable people, not just brains. Jeff Goldblum's performance, the uniqueness of Watson, and the dynamics between Franklin and the men in her shop are all part of the human side of science. Tarzza realizes that the leap from concrete models to formal inferences will be difficult for his students. So he reinforces their attempts by supporting different learning styles and using the affective value of the lesson.

What *shouldn't* we do with a movie in a Standards-based classroom?
- Emphasize the answer, rather than the process.
- Allow students to take in information passively.
- Show a video from start to finish without active analysis.
- Present the DNA model as a "finished product."

Resources for the Road

Biological Sciences Curriculum Study (BSCS). (1989). *Advances in Genetic Technology.* Lexington, MA: D. C. Heath.

Chargoff, E. (1950). Chemical Specificity of Nucleic Acids and Mechanisms of Their Enzymatic Degradation. *Experientia, 6,* 201–209.

Taylor, H. (1958). Organization and Duplication of Genetic Material. *Proceedings of the X International Congress on Genetics,* Vol. 1.

Watson, J. D., and Crick, F.H.C. (1953). Genetic Implications of the Structure of Deoxyribonucleic Acid. *Nature, 171,* 964–967.

Wilkins, M. (1956). Physical Studies of the Molecular Structure of DNA and Nucleoprotein. *Cold Spring Harbor, 21.*

Biological Evolution

Nature of the Learner

Most students will be able to sequence events over time and demonstrate an understanding of superposition and fossil evidence.

Practice with problem solving will help them understand the significance of mutations in populations and be able to make predictions from genetic data. As they explore, students will improve their ability to use statistics for making predictions and to apply their genetics understandings to evolution.

However, research has shown that most students who can pass tests about Darwinian evolution while they are in a biology class answer questions about evolution in a "Lamarckian" manner within a year after they leave their coursework. Emphasizing the quantitative foundation of evolutionary science and scrupulously avoiding careless language ("The butterfly can adapt to its surroundings by changing color...") can help students construct more permanent understandings.

History and Nature of Science

Historically, the science of evolution has progressed from observations to mathematical modeling and molecular biology. Students need to follow the same path from concrete to formal as they grow in logical reasoning ability. When studying natural selection, for example, students can find fossils in simulated rock strata (like Hutton), measure variations in populations (like Kettlewell or Gould), and then share today's data on the genomes of related organisms (found in almost any issue of *Science*).

Nature of Instruction

Since the time of Piaget, teachers have realized that students have difficulty transferring logical skills that they possess in one context (like mathematics class) to new environments (like the biology room down the hall). Teachers of evolutionary biology will discover that statistics must be taught, reinforced, and illustrated, and that students will need extensive practice using these techniques.

Fortunately, there is no limit to the examples of living things that can be measured to help students understand variation. The width of 500 dandelion stems; the breadth of a collection of brachiopod fossils; and the height, handspan, or reaction time of the students in the cafeteria are all convenient subjects for practice.

Most students come to class with the preconception that species are always discrete and constant, a belief partially supported by many classroom experiences dividing objects into discrete groups. Too much emphasis on taxonomy (as in the courses of the 1950s) can create the misconception that species are static, separate, and immutable. This serious misconception hinders students' progress in the study of modern evolution.

Assessment

Almost any activity that makes an effective lab can become a good authentic assessment. Consider

- burying fossils in layers of sand, shredded paper, or pet litter and asking students to develop a hypothetical evolutionary sequence from their understanding of the principle of superposition
- providing students with collections of shells of different length/depth ratios or paper insects of different color proportions from two hypothetical environments and asking students to measure, graph, and hypothesize about the environments in which they evolved

For more advanced students, data on the degree of

correspondence between the gene sequences of organisms in the same family (for example, felines) can be provided, and students can be asked to build a hypothetical evolutionary tree.

Personal and Social Perspectives

Biodiversity is an important concept that should be explored when studying evolution. This real-world application of evolutionary science helps students move from abstract to personal significance. A valuable analogy can be found in the preface of Paul Erlich's classic *Population Bomb:* A passenger in a airplane along a runway watches as, one by one, rivets are removed from the wing. Similarly, each individual species may seem unimportant, but taken together they create the stability of the ecosystem.

Science and Technology

Radioactive dating offers a quantitative basis for evolutionary science. Students can also practice using calipers and computer-interfaced measurement techniques in evolution labs, learning statistical techniques and degrees of accuracy.

A Classroom in Action

Jane Walker teaches both at-risk (sophomore) biology and Advanced Placement (senior) biology in a conservative community. She realizes that the Content Standards apply to both groups of students and faces the daily challenge of providing opportunities for all her students to master content.

Jane's students not only differ in ability, but many also have preconceptions about some content that present barriers to learning. Evolution is one such topic. Some students and their families are openly skeptical about this key concept. To find methods to help every student achieve this content standard, Jane first looks to the literature.

Research by Johnson and Peoples (1987) suggests that her students will accept the content of evolution more readily if they understand the nature of science. A second paper (Clough, 1994) suggests ways to reduce student hostility towards evolution, and a third (Scharmann, 1993) discusses how to design effective evolution instruction.

At both levels, Jane begins her unit on evolution by exploring student misconceptions. In small groups, students discuss what science is, how they would define evolution, and why some groups may oppose evolution education. By having students speculate rather than commit to a personal opinion, Jane creates a less threatening environment. Jane analyzes each misconception as it appears. Commonly, she hears
continued next page

Life Science   97

- Supernatural explanations are sufficient.
- Scientific theories are "guesses."
- Truth is determined by majority rule.
- Science can determine absolute truth.
- Areas of uncertainties should be rejected.

Jane knows that active learning produces the most sustainable learning. She presents the Sidney Harris cartoon (1977) that depicts two scientists at a chalkboard looking at a long series of mathematical calculations. Where a gap in the math appears, the board is labeled "Then a miracle occurs." The scientist in the foreground cautions, "I think you should be more explicit in step two." Students enjoy the cartoon, but Jane moves further by asking, "Why does the scientist want more information?" Eventually students agree that the supernatural explanation isn't very useful in science.

She explores the nature of "theory." Using constructivist dialog, Jane asks what makes something a law or theory, and she lists students' ideas on the board. She gives students a list of laws or theories; and in groups, they discuss how well their ideas fit the list. Students discover that their ideas are unsatisfactory. They are now more receptive to Jane's discussion of the relationships among hypotheses, theories, and laws.

The idea that science should be democratic recurs almost every year. Some students ask why the "alternative" theory of creationism shouldn't be given equal time. Jane must then return to a discovery approach to support students as they learn to distinguish science from public opinion. To give their unit a historical perspective, Jane has her students do independent research on Copernicus, who fought against the majority opinion of his time.

She asks students to analyze a common attitude, that early scientists were "inferior" because their ideas were later replaced as technology improved. Realizing that the notion of an Earth-centered universe dominated Western opinion for more than 2,000 years helps students understand that science will always be subject to public opinion. They eventually discover that science can be both evolutionary and revolutionary.

Good theories predict, explain, and provide a framework for future research (Moore, 1993). Jane uses examples from research to illustrate that incomplete data doesn't always result in a theory being abandoned. For example, punctuated equilibrium can be used to explain gaps in the fossil record without having to abandon natural selection.

In both of Jane's classes, many students are most attracted by the human element in science. She discusses Darwin's strengths and his human frailties, the effect of his work on society, and the Piltdown fraud of 1912. In the future, she plans to incorporate reenactment of historical debates, research of original papers, and for advanced students, more grounding in philosophy. Well along in incorporating the Standards into her teaching, Jane is continually exploring new paths.

Resources for the Road

Clough, Michael P. (1994, October). Diminish Students' Resistance to Biological Evolution. *The American Biology Teacher,* 56 (7), 409–415.

Johnson, Ronald L., and Peoples, E. Edward. (1987, February). The Role of Scientific Understanding in College: Student Acceptance of Evolution. *The American Biology Teacher,* 49 (2), 93–98.

Moore, John A. (1993). *Science As a Way of Knowing: The Foundations of Modern Biology.* Cambridge, MA: Harvard University Press.

Skehan, James W. (1986) Modren Science and the Book of Genesis, Arlington, Va: National Science Teachers Association.

Interdependence of Organisms

Nature of the Learner

Beginning biology students can usually understand niches and relate them to food chains. They can also predict the effects of one variable on a population and graph population data.

Through exploration and practice, students will learn to use data to predict changes over time, and they will understand the effects of several variables. The idea of a single variable being a limiting factor in a very complex ecosystem (such as nitrogen in a bog) will be difficult for most 9th and 10th graders without concrete exploration.

History and Nature of Science

Studying historic naturalists (like Audubon) can help students achieve affective goals (for example, caring about the environment), but the transition to a quantitative point of view may be more difficult.

There is a significant body of political opinion that holds that today's environmental science is not quantitative or "hard science." This position will have to be discussed and refuted for students to appreciate the implications of scientific data.

Today's environmental science relies more on quantity of data (and the worldwide communications that make large-scale studies possible) than on dramatic technologies or controlled experiments. This is the area of the Content Standards in which secondary students can often become real partners with researchers in studies of natural phenomena (for example, in cooperative projects such as a stream watch or a monitoring study of a migratory animal).

Nature of Instruction

Field studies are essential for students in ecology. Even inner-city schools have ample opportunities to observe niches of organisms. Pairing classroom or computer modeling with field studies can help students move from concrete to formal reasoning.

Creating model classroom environments (terraria, aquaria, stream tables) and using them to test hypotheses can help students understand the relationship between field and laboratory research. Many of the best instructional exercises in ecology involve extended studies of very small, simple communities: field quadrats, water quality studies of creeks or drains, or studies of animal feeders. Journals and portfolios are important communication tools for these studies.

Assessment

Depending on students' levels of cognitive development, assessing concepts can be done with real (pictorial, video, or actual) examples or with data sets. But we must be careful to separate the variables in assessments. Students can understand interdependence in real-life settings, but they may not be able to demonstrate the connections between statistics and inferences.

Videodiscs and CD-ROMs can provide good simulations of natural environments for group testing. Students will find it easier to reason about environments they can see and hear; they will find it more difficult to interpret data from tables and graphs. The ideal assessment will combine both forms of data so that students can learn to make connections and inferences from concrete and abstract examples.

Personal and Social Perspectives

Today's political climate associates environmental activism with emotional (non-quantitative) reasoning. Appropriate instruction can avoid that pitfall. Standards-based environmental education can lead naturally to environmental and political activism, and the best opportunities are often found in students' own communities.

The best environmental science models in schools today are studies, generated by students and their families, of local problems. Many have been supported and disseminated by the Toyota TAPESTRY award program. Examples include
- a study of the quality of well water in a county in southern Minnesota
- infrared photography of the animals in a threatened wetland area in the path of a proposed expressway
- weekly quantitative studies of refuse on a beach in Texas
- experiments on how to dispose of chicken wastes in Arkansas

Local problems provide the best environmental science curricula.

Science and Technology

Satellite photography produces the monitoring data for governments and environmental research scientists. This data is valuable for student discovery. GIS (Global Information Systems) software and satellite technology are now available for students in many classrooms across the nation.

Through the Internet, students are often contributing to important environmental science research. Project GLOBE (Global Learning and Observations To Benefit the Environment) at http://www.globe.gov can link your classroom to others.

Resource for the Road

Singletarry, Ted J., and Jordon, J. Richard. (1996, March). Exploring the GLOBE. *The Science Teacher,* 63 (3), 36–39.

A Classroom in Action

Each fall, thousands of students, their teachers, and other volunteers collaborate with scientists in a project that combines content knowledge in environmental science and technology with the fascinating life cycle of the monarch butterfly. How the monarch migrates each year to a small area of central Mexico from the northern United States is still one of the most intriguing mysteries of biology. And tracking this migration has become a great example of students using the communication tools of the 1990s to work together.

At Olathe High School, Brad Williamson began participating in the monarch project as a way to help students collaborate in investigating a scientific problem. (The Monarch Watch is now funded by the National Science Foundation, and the progress of student research across the nation can be followed on the World Wide Web.)

The Monarch Watch was once a collaboration among a few schools, but now thousands of classes chart the migration of the butterflies. Classes might begin by leaving an e-mail message to indicate their interest at monarch@falcon.cc.ukans.edu. If the class subscribes and becomes a member of the electronic mailing list/discussion group (DPLEX-L), they can post e-mail mes-

sages to and receive messages from all the subscribers who share their interest in the monarchs. Contact listserv@ukanaix.cc.ukans.edu for information on the discussion group. Participants can record data, exchange ideas, and pose questions about the migration research.

In this project, students learn that accurate, timely observations are an important part of environmental studies. They are often surprised that the validity of a scientific study may depend more on the quantity of data (number of observations) than on the sophistication of the method for gathering them. As students like Olathe's participate with university researchers in real investigations, their sense of the nature of science grows along with their excitement and interest.

Like many classes around the country, the life science students at Olathe also have a home page on the World Wide Web (http://129.237.246.134/) Visitors to the home page can read about migration and monarchs. Using the Internet, Olathe students share research with other students and participate in other schools' research projects.

Interdependence is an important concept in science—whether the subjects are butterflies, students, or researchers. The monarch project gives Olathe students a sense of the interdependence of the community of scientists as they become contributors to that community.

Matter, Energy, and Organization in Living Systems

Nature of the Learner

Beginning biology students can usually reason formally about biology content that they can see. They can follow directions in experiments and determine cause and effect in one-variable experiments.

But even with practice, students will have difficulty relating what they see to changes in molecules and using chemical knowledge to predict biological results. Because most of their knowledge about photosynthesis and respiration will have come from experience outside the classroom, students will hold on to many misconceptions during their study of these important topics.

History and Nature of Science

The historical experiments in photosynthesis (Van Helmont and Calvin) are ideal for helping students develop quantitative understanding in this content area. On the other hand, the classic experiments in nutrition and respiration, which have often been included in high school programs, are almost always inappropriate because they require unnecessary stress on organisms to provide predictable results.

Photosynthesis is the best area of biology in which to review the controlled experiment model for scientific research. However, we must be aware that students will still have difficulty separating the effects of multiple variables in classroom experiments.

Nature of Instruction

Photosynthesis and respiration are the most complex clusters of content objectives for beginning students. Appropriate instruction in these areas takes months and requires progressive and thorough laboratory exploration and continual assessment for misconceptions. This study must be closely associated with experiments in physical (inorganic) chemistry to help students relate what they know. Some schools have created integrated science programs to achieve this continuity; in other schools biology has been moved to the junior year.

Fortunately, the equipment and facilities for exploring photosynthesis are accessible to almost all classrooms and students. We will have the best results with designing units that are based on many laboratory experiences of gradually increasing complexity and by encouraging students to keep personal journals or other forms of continuous recordkeeping to chart their growth in understanding.

Assessment

When instruction is based on laboratories, assessment should be, too. Examples of such "tests" might include showing examples and asking questions. Given

- a geranium leaf with one section white because it has been covered, "Why is there no chlorophyll?" (photos or videos can substitute for real plants if necessary)
- a series of chromatography strips, "What is the relationship between the different proportions of pigments and the use of sunlight?"
- graphs of the metabolism of small reptiles over a range of temperatures, "Why do the temperatures change? What are the limits?"
- photos of a lawn with dandelions and grass during long spring days and shorter fall days, "How can you account for the differences you see?"

Personal and Social Perspectives

The study of respiration leads naturally to explorations of food chemistry, diet, and nutrition. Since growth and appearance are especially important to adolescents, these areas can be springboards for relevant lessons in biology.

The best Standards-based lessons in bioenergetics involve school activities. Consider taking over your school's weight room to explore metabolism. (Do the new, popular "power" drinks have ingredients that really raise metabolism?) Or move the classroom to the school cafeteria. (Could you lose weight if you only ate Type A lunches?)

Science and Technology

In this area a lot of laboratory experimentation can occur without a great deal of expensive equipment. Worldwide studies in photosynthesis (studies in deforestation, for example) rely on satellite technology. Students can access and download a variety of databases charting the changes in photosynthetic activity around the globe from a classroom computer and modem, giving them access to and a real understanding of the importance of telecommunications to botanists.

To help students understand the technologies that health care professionals use, consider a partnership with a local hospital that would allow students, for example, to measure metabolic rates.

A Classroom in Action

Maria Silva has taught biology for two years in a comprehensive high school where the majority of the science department has more than 15 years' experience. She wants to move her assessment methods toward the Standards but fears she will be pressured by her peers not to stir up "controversy." She also knows that students who do well on multiple-choice tests often fear authentic or performance tests, and that she may encounter parent resistance if she bases grades on them.

In her first year of teaching, Maria discovered that her students "survived" the unit on movement of materials in cells by memorizing the material. She wants to plan a unit that will promote greater understanding this year, and she contacts a biology teacher in another school for ideas. She establishes an e-mail relationship, and her teaching "coach" agrees to comment on her plans and her assessment results as the unit proceeds.

On the first day of the unit, students encounter board work when they enter the classroom—a quick question to be answered before Maria finishes the attendance report. But rather than ask for a definition or fact, Maria asks a version of a question that was validated by researcher Anton Lawson as a quick measure of formal reasoning. "Harry is four pennies tall, and his big brother Jim is six pennies tall. If you measure Harry in paper clips, he is six paper clips tall. How tall will Jim be in paper clips?" She knows that students who don't understand proportions will generally give the answer "eight," while those who do will answer "nine." But she is surprised to discover that the majority of her 15-year-old students answer "eight." So she reconsiders the pace of her instruction and adds simpler laboratories.

A key objective in Maria's unit is to have students understand the relationship between the size of a cell and its rate of absorption. She begins with several simple experiments that allow students to explore diffusion.
- First, she releases peppermint fragrance at one corner of the room and asks students to raise their hands as soon as they smell it. Then she asks students to draw what they believe to be the distribution of oil molecules in the air in the room. As she circles the room, she notes students who don't visualize air and fragrance molecules as separate particles and marks a seating chart to alert her to pay particular attention to those students at discussion time. The students put the drawings in their portfolios.
- She provides each student with a small "sausage" of dialysis tubing filled with starch water and a beaker with dilute iodine water. When students observe the iodine entering the cell model (but no starch leaving), she asks them to write a two-sentence explanation of what their observations tell them about the size of the pores in the tubing and the size of starch and iodine molecules.

Maria evaluates each answer for misconceptions and compares the answers to the picture-assessments she collected previously. (Later at home, she notes which students are able to demonstrate their understanding in pictures but not in words; she marks a code in the grade book about individual learning style.) The paragraphs are kept in portfolios, too.

For students who have difficulty understanding the dialysis-tubing experiment, Maria suggests a computer tutorial (Osmosis and Diffusion from EME Corporation, PO Box 2085, Danbury, CT 06813) to complete on their own. This tutorial allows students to see the movement of particles and also to change variables.

continued next page

As the subject matter becomes more difficult, Maria injects some humor and a "real-life" problem into her unit. She shows five minutes of the classic science fiction video *The Blob* and asks student groups to generate a scientific explanation of why the blob couldn't exist. She puts their trial explanations in their portfolios.

For a major experiment, Maria asks student groups to determine how the size of a cell affects the ability of its membrane to absorb nutrients passively. She provides each group with a potato and asks students to cut cubes at 0.5 cm, 1.0 cm, and 1.5 cm on a side. At a signal, each group drops the cubes into dilute iodine solution for 30 minutes. The cubes are then removed, the surfaces are dried, and the cubes are split.

For each cube students determine the original volume, the volume of the cube remaining white, the volume that turned black, and the percent that turned black. They enter this data on a spreadsheet.

Maria knows from her first-day assessment that some of her students will not understand the process of percent and proportionality. As the experiment proceeds, she moves from group to group asking the less logically minded students to explain the mathematics of determining the percent absorbed. Some make progress through this verbal interaction. A few are still confused; so Maria notes for their math teacher that some integrated review is needed. She also contacts the teacher consultant for her two special education students and gives the consultant a complete set of unit plans and objectives.

With the unit completed, Maria must design a test. Her school community expects a final test grade even though she has great doubt about her ability to measure validly the many variables she has observed in class. She revisits the objectives she had designed for the unit:
- skill in observing results and recording data
- an understanding of the particle theory of matter
- use of proportional reasoning (in predicting the relationship between cell size and movement through a membrane)
- use of scientific reasoning (in understanding the relationship between experimental in vitro results and in vivo phenomena)
- achievement at the application level— students being able to apply what they have learned to real-life situations
- growth in students' abilities to analyze their own learning and plan for future growth

Maria's final test will take two days and involve five clusters of evaluation tasks. The first three parts are paper-and-pencil tests.

Part I asks students to review what they observed during the instructional experiences. Maria is confident that well-designed objective items with a low reading level will be appropriate, but she makes arrangements for her special education students to have help reading the questions.

Part II is a series of open-ended questions that allow students to draw or write sentences to explain situations Maria describes in pictures and words.

Part III asks students to look at the setup of the dialysis tubing experiment they did and to predict (by choosing among options) the most likely missing data points.

Parts IV and V involve real equipment. In Part IV, students may review and examine previous lab setups and use real equipment to demonstrate their own reasoning. For example, one station asks students to design an experiment to determine if
continued next page

104   Life Science

iodine diffuses through dialysis tubing more quickly if the temperature is higher. A variety of materials are available for students to choose from as they write their own experimental procedure.

Part V asks students to apply what they know to real-life situations. After watching a video of a *Paramecium* exposed to waters of three different salinities, students must offer explanations of why the contractile vacuole's movement changes. They are also asked to explain why soft contact lenses feel uncomfortable when placed in the eye if they are stored in pure water rather than in saline solution.

Finally, Maria asks her students to reexamine the drawings in their portfolios and to write a paragraph about how their ideas have changed. This section isn't graded but allows her to plan for the next unit and for next year. Maria carefully compares these evaluations to previous test scores and to the scores on the last parts of the assessment to determine how internally consistent her test is.

Behavior of Organisms

Nature of the Learner

Beginning biology students can usually observe qualitative behaviors, record observations clearly, and interpret graphs.

Even after practice, students will have difficulty making quantitative observations of behavior and converting these observations to graphic format. Only a few students will make enough progress to predict behaviors from data or to relate them to feedback systems. Likewise, most will have difficulty relating behaviors to evolutionary needs and relating endocrine changes to environmental stress.

Learners at this level are particularly fascinated, and often confused, by issues related to individual and community health. While we should take advantage of the motivation in this content area, we should be careful not to create misconceptions or unreasonable fears.

History and Nature of Science

Ethology (the science of animal behavior) will challenge students' preconceptions about how science happens. Many students will relate only to controlled experimentation. Review and discussion of the *quantitative* aspects of the work of Goodall, Fosse, and Rumbaugh can be a good bridge from casual observations of behavior to modern science. Teaching students to quantify observations of living things can teach not only a new "scientific method" but also patience and accuracy.

Nature of Instruction

Students can learn the techniques of ethology by studying invertebrates in the classroom. Initially, they can observe differences within and among species. After becoming familiar with the normal responses ("taxes") of cultured invertebrates, students can introduce nonlethal variables to the organisms' environment. We must constantly emphasize the quantitative aspects of the methods in order to avoid the impression that just "liking animals" is biology.

Many teachers encourage students to raise and become familiar with the normal behaviors of invertebrate cultures (mealworms, *Daphnia*, tubifex worms). In other classrooms, students choose organisms to observe and write about in their journals over long periods (pets, fish, backyard birds, or younger siblings).

Assessment

Authentic assessments can challenge students to design their own experiments on topics such as "Do *Planaria* prefer light or darkness?" or "What fraction of my cat's time is spent being active?"

CD-ROM videos of animal behavior or plant tropisms can become the basis for assessments in which students are asked to interpret, relate, or apply concepts they have learned to new situations.

Personal and Social Perspectives

Studying the behavior of organisms leads naturally to encouraging respect for organisms and humane treatment of animals. These objectives can become part of any unit in ethology without anthropomorphizing animals.

This standard is an ideal context in which to discuss the relationship of behaviors to Personal and Community Health. Helping students relate their own behaviors (and those of others in their community) to health statistics and consequences is an important outcome.

Science and Technology

In ethology, students can improve their ability to use statistical techniques, computerized graphing, and data analysis. The kinds of analyses in animal behavior will be different from those in controlled experiments, and they will expand students' capacity to use computers as tools.

Using video technology can help students sharpen their observational powers and refine their ability to communicate understanding to one another.

A Classroom in Action

Stanley Koski has taught in a rural Michigan school for 15 years. His school survives on one of the lowest per-student fundings in the state (his science budget has been less than $3 per student for many years).

Despite Stan's limited budget, he has always managed to provide hands-on experiences for his life science students. About 40 percent of the textbook, which was new when Stan arrived, is devoted to comparative phylogeny. So Stan has usually provided a dissection specimen for each major phylum, and students have worked in groups to study the anatomy of each phylum.

At the state science teachers meeting, Stan goes to two workshops—one on the state's core curriculum and one on the National Science Education Standards. Neither document gives much emphasis to comparative phylogeny. While dissection is not specifically discouraged, the content it had been used to illustrate isn't considered a high priority anymore. Behavior, an area Stan has never formally studied, is emphasized.

At a break in the meeting he talks with other teachers. "What do you do for lab investigations in the area of behavior?" he asks. Thinking he will get a laugh, he adds, "for little or no cost."

Mary Brown doesn't even smile; she offers Stan a handout from the workshop she has just given: "Bait Shop Biology." She explains that her inner-city high school has as limited a budget as his rural school, and that she has had many successes by asking students to explore animal behavior with invertebrates.

In her classes, each student group is given a culture of inexpensive invertebrates, including mealworms, earth and red worms, mousies, crickets, vinegar eels, fruitflies, and *Daphnia* (which she has maintained herself over the years). Because of the red tape at Mary's school, she doesn't even try to get specimens through a supply house; "I just stop at the bait shop and the pet shop," she explains.

continued next page

Life Science

Before the organisms are completely entrusted to the group's care, students must demonstrate that they have researched good culture techniques and that they are prepared to keep the organisms alive and healthy. After about two weeks of successful culturing, Mary asks the groups to think of a single variable to explore with their cultures. They might determine how the specimens respond to light, vibration, heat, pH (for those in liquid environments), or magnetism. No variable is allowed that would kill or harm the organisms.

To create their experiments, students may use a number of inexpensive tools. Disposable plastic tubes from thermometers make good "tunnels" for fruitflies when determining whether they like light or darkness. For *Daphnia* and vinegar eels, strips of typon tubing from the hardware store work well. Students create trails and mazes for worms and crickets. They are asked to write up the labs with purpose, procedure, and data tables of results. One or a few trials are never sufficient.

Later in the behavior unit, Mary introduces some techniques of quantitative observation. She instructs students to find an organism they will be able to observe undisturbed for at least 10 15-minute periods. Choices include birds at a feeder, pets, squirrels, aquarium fish, and even younger siblings. Students keep a log of the percent of time organisms spent in specific activities. For example, the categories for a pet dog include sleeping, begging, playing, exploring, searching, eating, and going outside.

Students describe specific behaviors in one-page paragraphs and later research the behaviors to determine their survival value (for example, dogs circle before they sleep to pat down grass; kittens knead before they suckle to work milk from the mother cat). Class discussions of invertebrate behaviors inevitably lead to speculations about the survival value of human behaviors.

The informal discussion at the convention led Stan to do more research. He wasn't confident he could change his curriculum overnight, so he asked Mary if she would become a "phone mentor" for a few weeks in the spring. She went further and volunteered to share her class's data on pet behavior with Stan's class via the Internet. Stan couldn't promise his school would be online by that time, but he said he'd try.

The convention stimulated both awareness and some concern in Stan. He feared that the search for an inquiry-rich program to match the new Standards would be complicated and expensive. He didn't have all the answers or even enough for a semester. But with the encouragement of Mary and others at the meeting, he was certainly ready to take the trip to the bait shop for his next lab.

Resource for the Road

Texley, Juliana. (1993, May). Bait-Shop Biology. *The Science Teacher,* 60 (5), 23–25.

A Classroom in Action

Integrating Life Science and Other Subjects

Kathy Liu and other teachers at Serramonte High School developed a series of interdisciplinary projects that integrate science, social studies, language arts, and mathematics to support a restructured secondary curriculum. The Nutrition Project illustrates active learning and addresses the content areas of Behavior and Personal and Community Health.

In developing, implementing, and evaluating this interdisciplinary program, the planning team relied on *Benchmarks for Science Literacy* from Project 2061 (1993) and state outcomes to guide their move toward the Standards. Here are examples of tasks in a six-week unit on nutrition: Create appropriate menus for Americans today and those living in the 1930s. Or, identify a nutritional problem, propose a solution, and select an appropriate audience for a presentation. Plan the project using a Gannt chart. Acquire the background information to complete the project by reading literature, practicing necessary algebra skills, reviewing chemical formula notation, tracing the process of digestion, and studying the economics of food production, distribution, and marketing. Demonstrate mastery of background knowledge by synthesizing it into an action plan, by successfully completing an individual written or verbal test, or by writing an expository or persuasive paper.

For each part of the unit, the team involved students in planning, implementing, and evaluating their own achievements. This included student help on designing grading rubrics. To provide a literary base, students read *Black Boy;* so the grading rubric might look like this:

- Level 1 response—1 page
- Level 2 response—2 pages
- Level 3 response—3 pages
- Honors—Figure out a typical day's menu for Richard. Analyze the menu for its nutritional content, and evaluate it for Richard's health. Or eat the food on the menu for three days, and report on how it felt to do that.

Similar rubrics were developed for content mastery of mathematics, chemistry, nutrition, food composition, limiting reagents and energy, the digestive system, and American history. The units also included work on economics (supply, demand, and price of foods), types of market in the economy, and the role of government in the economy.

Students were asked to develop an action-plan pyramid to help them see how knowledge and analysis must precede action and decision-making. It also helped students stay on task and encouraged their involvement in planning their own learning.

The problems students investigated involved the personal and social perspectives of the science they had studied. The course outline framed each sample problem as a hypothesis: "People are malnourished because they do not know how they should eat." "People are malnourished because they choose the wrong food." "People are malnourished because they cannot afford nutritious food." The students considered their audience (parents, homeless advocates, other students) and planned a presentation for that audience.

The Serramonte High School interdisciplinary projects demonstrated that developing integrated units is not only good for students but also for teachers. Such an effort encourages *everyone* to collaborate.

Resource for the Road

Liu, Kathy. (1995, October). Rubrics Revisited. *The Science Teacher, 62* (7), 49–51.

Earth and Space Science

Many people have misconceptions about the fundamentals of Earth and space science. In popular culture, cartoon dinosaurs and humans romp together, Earth is a few thousand years old, and winged planes roar through space. What is even more unfortunate is that most citizens lack the scientific background to understand the many Earth science-related environmental issues. To deal with global warming, groundwater pollution, ozone depletion, or dwindling natural resources, our schools must reexamine their Earth and space science programs in the context of the Standards.

Earth science is not just the study of geology or rocks. It involves integrating many sciences to achieve the core content outlined for the secondary level:
- *Energy in the Earth System.* Energy on Earth comes from two sources: the Sun and Earth's internal radioactivity.
- *Geochemical Cycles.* The Earth system is composed of interacting subsystems of the geosphere, hydrosphere, atmosphere, and biosphere.
- *Origin and Evolution of the Earth System.* Earth is more than 4.6 billion years old and still evolving.
- *Origin and Evolution of the Universe.* The universe is a vast, evolving system; the physical laws relating matter with energy can be used to test theories about its origin.

The Content Standards related to Science in Personal and Social Perspectives are a vital part of the study of Earth science. Particularly relevant to secondary students are
- *Environmental Quality.* Earth is being altered by human activity.
- *Science and Technology in Local, National, and Global Challenges.* The use of scientific thinking and technology will increase humankind's understanding of Earth and allow us to use its resources better.

A Standards-based course in Earth science will include many concepts in astronomy, meteorology, hydrology, oceanography, and geology. Teaching about the Earth system is a good way to demonstrate how principles of chemistry, physics, and biology have real-world applications that affect students personally.

What is exciting to the Earth science community is that the importance of the discipline's content knowledge to secondary students is finally being recognized as an equal partner with the physical and life sciences. Many professional organizations have worked for many years to ensure this outcome. Earth science content is included in Project 2061's *Benchmarks for Science Literacy;* in NSTA's Scope, Sequence, and Coordination of Secondary School Science project's *Content Core* and A *High School Framework for National Science Education Standards*; and now in the *National Science Education Standards.*

By its nature, Earth science lends itself to using examples from students' lives. The school grounds or neighborhood often offer opportunities for hands-on activities,

and current events such as storms, earthquakes, and volcanic eruptions can become a dynamic part of daily lessons. Good teachers build their programs on issues close to home, such as the community water supply, the location of the nearest landfill, or the effects of mining or quarrying.

But many concepts of Earth and space science are abstract and difficult. To move from local and contemporary examples to the vast time scales of mountain building, the age of the universe, or the vast distances of plate movement or space science, lessons will have to be carefully designed. Beginning with concrete examples and slowly bridging to the abstract—through computer simulations, videos, slides, data gathering and observation—can help students gain new skills in logic.

Assessment in tomorrow's Earth and space science classroom will go beyond the traditional multiple-choice test; it will ask students to demonstrate important, integrated concepts by solving a problem or interpreting new data. In Earth science, students will learn approaches to historical and empirical research that are different from the traditional methods used by the physical and life scientists they have studied.

Choosing a pathway toward the Standards will require significant changes in courses and programs for many schools. Some will use an integrated program based on Unifying Concepts and Processes (such as Systems, Order, and Organization; Constancy, Change, and Measurement) instead of teaching a separate course in Earth science. Others may change the course sequence to place the naturally integrated curricula of Earth and space science after beginning courses in physical and life sciences. Many sample programs and curricula are described in the following pages.

One thing is certain: An Earth science curriculum moving toward the Standards will have very few lectures. We will be coaches and facilitators rather than deliverers of knowledge. Our students will learn Earth science by doing, not by being told. Instruction will begin by identifying misconceptions, providing concrete experiences, and having students observe, collect, and interpret data. Students will work often in groups.

The classroom for tomorrow's Earth and space science program will be different, too. Finding pertinent resources will be a priority. There is a virtual explosion of information on geology, environmental issues, local and global climate, and planetary and space sciences. Videodiscs and CD-ROMs provide volumes of information at low cost, and we and our students can learn to find the data we need in them.

The next step is to access data online. The Internet, commercial online services, and even satellite downlinks are becoming essential components of the Standards-based Earth and space science classroom. When studying weather, it's best to download current information and use it to plot graphs, analyze weather patterns, and make local predictions, instead of using old information. Following the latest NASA launch or the work of the Hubble telescope is more likely to attract future astronomers than reading historical data in a text. Students can also access databases about environmental concerns such as local carbon dioxide or air and water pollution levels.

To teach a Standards-based course in Earth and space science, we must have the support of our school and community. Opportunities for field trips, for example, are essential. Firsthand observations in a stream are much more effective than using a stream table. Field trips give students a chance to do real science; even a camp-out in the school parking lot can provide a memorable view of the night sky. Resource persons from the community and personnel from government agencies involved in surveying, planning, agriculture, and natural resources can also bring real-world issues to the classroom.

Students can become partners in doing science by joining projects such as GLOBE (Global Learning and Observations to Benefit the Environment) available on the World Wide Web at http://www.globe.gov (Singletarry, 1996). By doing such projects, students can collect data and make invaluable contributions to science today.

To keep up with the knowledge explosion in Earth and space science, we will need help. Professional organizations provide valuable links to our colleagues—having a network of peers is a resource beyond measure. By attending conferences and workshops, we will meet colleagues with whom to network, keep current with new ideas, and most important, keep our enthusiasm high.

Moving toward the Standards will be both exciting and frustrating. Paul DeHart Hurd often tells us, "Everyone is in favor of progress but no one wants to change." As Earth and space science teachers, we will need to focus on major goals for our students. We will have to consider carefully why we are covering certain concepts and remember that "less is more." Our goals should include not only developing basic concepts but also supporting students' abilities to make personal decisions about how to use Earth's resources. Through Earth and space science we can encourage students to respond appropriately to societal issues and become lifelong learners.

Resource for the Road

Singletarry, Ted J., and Jordan, J. Richard. (1996, March). Exploring the GLOBE. *The Science Teacher,* 63 (3), 36–39.

Energy in the Earth System

Nature of the Learner

The essential concepts of energy in Earth science require formal reasoning in a physical science context. Only about half of ninth-grade students can demonstrate this level of understanding on paper-and-pencil assessments; but after exploring and experimenting, many more students will be able to discuss concepts or demonstrate them through concrete responses.

The content area includes concepts that apply to small systems and by extension to much large systems. Ninth-grade students will understand concepts more easily in the context of local examples; only a few will be able to extend this understanding to the global system without assistance.

History and Nature of Science

Thermodynamic principles underlie this content area (but will be new to most students). The ability of meteorologists to understand and make predictions from climate data has increased with the increased sensitivity of their instruments and their ability to integrate data from diverse sources. This is a wonderful example (and there are many more) to illustrate the close relationship between technology and science.

The internal and external sources of energy that drive Earth systems are important in this content area. However, most students will come to the study of Earth science with the misconception that the Sun is the only relevant source of energy. The historical development of Wegener's hypothesis of continental drift, the skepticism of the scientific community, and the eventual accumulation of evidence that renewed interest in the subject illustrate both content and inquiry methods to students. Emphasizing "how we know" and what patterns of logic have been used by geologists in the past can be valuable today.

Nature of Instruction

In the Standards-based classroom, each concept should be illustrated by student experimentation, followed by data collection in the local community, before students are encouraged to extend their understandings to the Earth system. Where classroom experiments are not possible (such as in tectonics), data from real-world situations (such as earthquakes) provide relevant opportunities for discovery.

Earth and space science instruction must always use the community as a resource. Lack of travel funds should not prevent classes from exploring the geology of the schoolyard or the nearest construction site. Even the materials from which the school was built can be studied in Earth science.

Assessment

Because students will not be able to demonstrate formal understanding of Earth energy systems, assessments that incorporate maps, models, or visual examples will yield more valid data.

Some of students' most common misconceptions involve climate. Good questioning can become a constant assessment in the class setting; this keeps us from moving too quickly or skipping over opportunities to confront and dispel erroneous misconceptions.

Personal and Social Perspectives

The ability of Earth scientists to understand many natural disasters is directly related to their ability to measure energy in Earth systems. Consider these examples:
- Earthquake predictions depend on understanding the role of energy in plate tectonics.
- Hurricane movements are predicted by understanding energy exchanges in global climate patterns.
- Scientists are very interested in understanding the role of energy in global warming.

The study of Energy in the Earth's System can begin and end with real-world situations which make the difficult, abstract concepts of energy transfer more personal to high school students. Start with headlines or news clips. Students can extend what they read by composing "the rest of the story"—the science behind the event.

Science and Technology

Working models of seismographs, weather data-collecting equipment, and satellite receivers can be constructed in classroom settings—even in schools with small budgets. Building a measuring device, even if its sophistication is low, helps students appreciate the quantitative nature of science and the relationship between technology and scientific progress.

Where budgets are limited, we might consider a partnership with a local airport or marina. Good weather data is available by phone (FAA) and on cable television as well.

A Classroom in Action

Rachel Stein knows that most ninth graders have persistent misconceptions about Earth science, and that her students will only be able to reason formally about concepts they have explored concretely. To move toward the Standards, Rachel develops a unit on weather and climate that incorporates many different experiences.

Clouds are certainly familiar to her students. She begins the unit by asking students to observe and discuss clouds, and she probes for potential misconceptions that will be challenged through later classroom experiences.

Students then use a two-liter plastic bottle to make a model of a cloud. She obtains liquid crystal temperature strips at an aquarium store and asks students to place a strip, a small amount of water, and smoke from a match in the bottle before capping it. By squeezing and releasing the bottle, students can discover how temperature and pressure relate.

Clouds move across Earth's surface propelled by the energy of temperature differentials. Rachel asks teams of students to research, design, and present to the class a demonstration to illustrate the effects of solar energy on weather. The cooperative project gives students an opportunity to discuss their developing ideas in a nonthreatening environment.

From the lab, students move to technology. There they access weather data over the Internet, from the Weather Channel, or from a satellite receiving station at school. They chart the data they receive and relate it to personal observations of local changes in weather, pressure, humidity, and frontal systems.

In Rachel's classroom, text readings and verbal explanations are carefully interspersed among hands-on experiences to avoid asking students to accept explanations without question. Videos are also used, when appropriate, to help students interpret their own experiences. The video series *Everyday Weather* (American Meteorological Society) has been valuable, but students are always encouraged to view programs "actively"—at key points, Rachel stops the program for questioning, demonstrations, or experiments.

Assessment follows instruction. Rachel gives students weather data they have not seen before and asks them to interpret it and to make next-day forecasts for a mystery area. Because Rachel does not wait until the end of a week or unit to assess understanding, she often finds that she must double back and repeat a concept students have misunderstood.

Rachel knows she must update her own skills continually to meet the Standards. Some sources of new ideas appear below.

Resources for the Road

American Meteorological Society, Educational Program, 1701 K Street NW, Suite 300, Washington, DC 20006-1509.

National Earth Science Teachers Association (NESTA), PO Box 53213, Washington, DC 20009.

National Center for Atmospheric Research, PO Box 3000, Boulder CO 80307-3000 (slides, videos, and articles).

Williams, Jack. (1992). *USA Today, The Weather Book.* New York: Vintage Books.

Geochemical Cycles

Nature of the Learner

By the secondary level, students will be able to understand the conservation of mass and the conservation of elements in Earth systems with consistency. However, their ability to extend this logic to new areas will be inconsistent.

They will have difficulty following concepts of the conservation of matter in chemical changes and the application of conservation concepts during change of state. Despite instruction, many students leave secondary school with unreliable understanding of the conservation of matter and energy.

History and Nature of Science

Earth is a closed system. That concept underlies every principle in the content area of geochemical cycles. Scientists depend on sensitive, long-term data-collection techniques to demonstrate the cycling of materials through Earth's system. Because historic advances in this area have depended on data collected by many teams of researchers, the study of geochemical cycles will introduce students to the need for international collaboration in the scientific community. Many Internet-based projects allow students to become real partners in this collaboration.

Nature of Instruction

Studying small, closed ecosystems can help students understand Earth as a closed system. Carbon may be the most easily traced element in such a system; but calcium, oxygen, and water cycles can also be demonstrated in laboratory situations. As we move toward the Standards, we will devote longer periods of time to the study of geochemical cycles. Since ideas related to these cycles require mature reasoning in high school students, we will need to embed authentic assessments in programs each day—questioning, drawing, group discussions, and journal entries.

To construct understanding about Earth cycles, students need a wide variety of knowledge and experience. Some examples include
- observing many rock samples and simulating the rock cycle with crayons or clay
- observing the effects of solar energy on samples of atmospheric gases
- simulating the hydrologic cycle in a closed container
- using field data (from petroleum exploration, for example) to plot world reserves

Assessment

Assessments of students' learning about small-scale cycles are common in classrooms today, but valid assessments of students' knowledge of large-scale geochemical cycles will require access to data, visuals, or examples to make them relevant and to make the assessment a valid measure of student understanding.

We might assess students' ability to apply concepts by asking them to
- relate the occurrence of limestone or shale deposits to the geology of the past
- work as teams to research the origin of the school's building materials
- analyze data from test wells to examine the extent of penetration of a pesticide or fertilizer on a farm
- trace the source of a pollutant or the change in pH in a stream

Personal and Social Perspectives

Once students understand cycles in Earth systems, they can transfer that knowledge to relevant social problems such as
- the persistence of pollutants (like pesticides) in the environment
- the need to recycle limited resources (like aluminum)
- decisionmaking on the use of nonrenewable

versus renewable resources
- determination of the source of the school's water supply or study of the source and quality of a local stream
- the results of global warming or the rise of sea level on a particular area

Many classes have turned to senior citizens as resources when studying climatic cycles in their community. ("Where was the shoreline when you were young?" "What year had the worst storms?") Senior citizens can also help students understand the effects of development on water quality in the community.

Students should reach out to other areas as their understanding expands. They could
- set up a monitoring system for a local river and develop a communications network to compare their data with data from a student group working upstream or downstream
- accompany a local well-driller, if possible, to look for hard or soft water

Science and Technology

The sensitivity of measuring instruments (often to parts-per-billion) is a major factor in the growth of this area of Earth science. Students should have the opportunity to explore the reliability of various measuring devices and to compare the sensitivity of various technologies in locating resources and detecting pollutants. Many schools have video production facilities which can provide the means to develop productions about the geology of the community.

Since very few schools will have access to the latest technologies, partnerships with government or businesses can be exciting. Consider sending student-collected samples to a testing agency to get good data. (A field trip to the agency would be ideal.)

A Classroom in Action

In a high school Regents Earth science class in New York, Jorge Cordoba introduced a unit on the water cycle by discussing local water supply and water pollution problems. He used newspaper articles on lead pollution in the local schools, recent droughts, restrictions on water use, and turf battles over water supplies in upstate New York.

A few days into the unit a student, Alyssa DiBello, mentioned that the residents of her trailer park were told not to use the water. Since water came out of the faucets, Alyssa wondered why she couldn't wash her hair with it or drink it. After a brief discussion, the class decided they wanted to tackle this problem. Jorge agreed.

That evening an article appeared in the paper about the water pollution problem at the trailer park. Alyssa's parents had talked to the manager of the trailer park and discovered that during a routine test of the well water, gasoline was found at levels considered unhealthy for human consumption. When Jorge asked students where the contamination in the well

might have come from, he found that many students didn't know the relationships among local precipitation, evapotranspiration, water table levels, and availability of water in wells.

continued next page

The class decided that they not only wanted to understand the water problem at the trailer park, but they also wanted to know about the source of water for each of their homes. They plotted the location of each home on a topographic map and found that the school district, a combined urban, suburban, and rural community, had many sources of water. These included a city water supply, two town water supplies, individual wells (including a few shallow, hand-dug ones), and groups of residences drawing from one well.

The class was divided into five study groups: one for the trailer park gasoline pollution problem, one for the city water supply, two for the town water supplies, and one for well water. Over the next few days, each group was to give a short oral report to the class, provide a written report, and display their findings in the hall. The students made phone calls, visited the headquarters of the municipal water suppliers, researched the gasoline pollution problem, and found out the sources and supply of well water.

One of the students had a neighbor who was a well driller, and he was asked to come to class and discuss his job. After this talk, some of the students visited a drilling site and reported back to the rest of the class. They discovered that the well driller had a great deal of data about the geology of their community in his log.

In addition to learning firsthand from the driller, the well group located U.S. Geological Survey reports on local groundwater conditions and wells. Hydrology and geology books provided the gasoline pollution group with background information; students also obtained information from the state's environmental resources department. They found that pipe connections to underground gasoline tanks often broke or separated due to frost, allowing gasoline to infiltrate the water table, from which it could flow into the trailer park.

The groups studying the town and city water supplies found that water came from the local reservoirs, which were supplied by local streams, which were dependent on local precipitation. These supplies were backed up by taps into the aqueducts that supplied New York City with water. Having the security of the backup carried the price of living with restrictions on water use in times of drought (for example, limits on watering lawns and washing cars).

Students discovered that a new soil incineration plant would be built near the school and that this could solve the gasoline pollution problem. One student noted that this plant might create other pollution problems in the atmosphere, on land, and in the Hudson River, which is adjacent to the plant. In the end, the water supply problems of the trailer park were solved by an expensive connection to one of the town water supplies.

Interest in the water supply problem didn't end with the unit test. Jorge read about a 1992 TAPESTRY award-winning project in Minnesota, in which a class mapped and tested wells in their farming community west of Rochester. Jorge called the teacher, Tim Johnson, and Jorge's class contacted the Minnesota class via the Internet and shared data. (For more information on Johnson's project, "Groundwater Sensitivity to Pollution in Minnesota," contact him at Kasson-Mantorville High School, PO Box 158, Kasson, MN 55944.)

Jorge's students learned much more than the water cycle in this unit, which they helped design, than students had in previous years. The personal and social perspectives of their content knowledge became the motivating factors for research, and the solutions they explored in a local problem had applications in other areas of the community. Students achieved Content Standards in Geochemical Cycles, Science As Inquiry, and Science in Personal and Social Perspectives. At the same time, they mastered research techniques that will help them become lifelong learners.

Origin and Evolution of the Earth System

Nature of the Learner

Most ninth-grade Earth science students will be able to build physical models and transfer conclusions based on them to real-world examples, such as sequencing fossils, describing erosion, and observing compaction. They will also be able to sequence chronological events, but their ability to comprehend and make inferences from the vast scale of geologic time will develop only with time and exploration.

Frequently students have misconceptions about
- the great age of the Earth
- the position of stars relative to other objects in our solar system
- the motion of plates, planets, and stars

Misconceptions about the nature of science, a "young Earth," or the validity of the geologic record persist in many students despite instruction. The media reinforces the natural tendency of adolescents to resort to "supernatural" or "magical" explanations for phenomena that are difficult to understand.

History and Nature of Science

In the past 200 years, the study of Earth's evolution has moved from an observational science to a quantitative one; students can follow the same path of logical growth by reading and studying the work of Hutton, Smith, Wegener, and Gould as they explore concrete models of geology.

As they learn content, students should develop the ability to use precision measuring instruments and to collect, combine, and analyze large quantities of data. Using computers to compile statistics can help students understand why learning about Earth's evolution has accelerated in this century.

Earth's origin is a key area in which students should develop a sound understanding of the nature of scientific theory. Using a historical approach to the body of evidence that established our current model of the solar system, students can gradually construct a personal understanding of the strength of "theory" that can be extended to the current scientific view of Earth's history. Because of persistent misconceptions about scientific theory in this content area, attention to the nature of science is crucial. Many students will equate uncertainty with untruth.

Nature of Instruction

Here, as elsewhere, the level of content understanding that can be achieved depends on the age of the student. In most school systems, Earth science is taught in eighth or ninth grade. At this level most students lack the knowledge of statistics and physical science to appreciate some of the most recent developments in the field. So instruction must proceed very slowly, or we have to teach some simple statistics and physical science concepts. Because of this difficulty, many new Standards-based curricula offer Earth science at grade 11 or 12.

Whatever the maturity of the students, Earth science instruction should follow the learning cycle:
- real-world observation
- generation of questions
- data collection
- physical modeling
- application and extension

Earth and Space Science   117

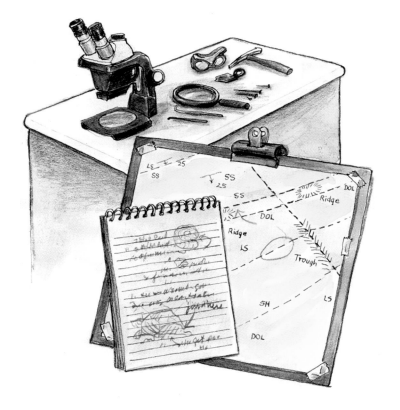

When the scale of real-world examples exceeds the size of the classroom, models are invaluable. Here are some examples:
- Time can be modeled using paper strips, dots, fences, or walls.
- Meteorite impacts can be modeled in plaster or sand.
- Stream erosion can be modeled in lab settings and studied with photo data.
- Earthquake foci can be plotted, graphed, and analyzed.

Assessment

Like instruction, authentic assessment should involve interpretation of real-world data such as photos, topographic maps, heat readings, or tectonic data. Students could be asked to view photos of the Moon and identify Earth-like processes. We might give students satellite images of a flood and ask them to sequence the images chronologically.

Students' understanding of the processes used by astronomers can be assessed through group assignments:
- Using satellite images of moons of other planets, determine the kinds of information that can be learned from crater studies.
- Using satellite images of the Mississippi river before and after the 1993 flood, list the information that can be extracted about the scope of the flood.

To assess understanding of the mathematics of scale; have student groups use string (1 cm = 1 million km) to demonstrate the distance of the Oort cloud of comets from Earth.

Personal and Social Perspectives

Students are usually surprised and motivated by the discovery of geologic phenomena in their own community. Evidence of glaciers or the remnants of an ancient flood or a meteor crater can go unrecognized for many years until seen through a geologist's eyes. Once students begin to see the world as geologists, they will maintain this habit throughout their lives, enriching their travel and leisure.

Science and Technology

Satellite photography is both a classroom tool and a tool used by scientists. Extensive use of such data can put students in touch with up-to-the-minute research. Data is available through direct satellite download or through the World Wide Web. Every Earth science class should incorporate extensive satellite data. Consider contacting a local university to get satellite data about the school grounds.

A Classroom in Action

Anna Kim was glad that her home state of New York was geologically exciting—where billion-year-old rock meets late Pleistocene (Ice Age) sediments. But the vastness of geologic time still left many of her students confused.

In 30 years of teaching, many of Anna's students had brought in rocks and fossils. This year, before starting a unit on Earth history, she decided to formalize the collection process. She asked each student to find a fossil in the local area. Specimens were labeled and placed in low-sided cardboard boxes. She left these boxes on a side counter near the door so students would see them on entering the room. Each fossil was numbered and an index card was kept for each specimen in a file box.

The fossils generated a number of questions: "What is it?" "How can we find out how old it is?" "What present-day life form is it related to?" "Why don't the life forms look like the present-day life forms of our area?" A few students understood the capacity of glaciers to carry material great distances; they suggested that the fossils collected in the neighborhood could have come from almost anywhere north of their locale.

Anna understood that student-directed inquiry would be the best approach to lead students to a real understanding of Earth history. To answer the questions they had generated, the class was divided into groups and given The Golden Nature Guide *Fossils: A Guide to Prehistoric Life* and various college reference books on paleontology and historical geology. Each group was assigned the task of studying a group of similar fossils in the class collection and determining what kind of life form the fossils represented, what environment they lived in, what information they might add to the existing knowledge of the history and evolution of the area, and what living relatives exist today on Earth. (Anna salted the collection of student fossils with some specimens of her own so there would be a more complete representation of local fossils.) Students' discoveries were to be added to the index card for each specimen.

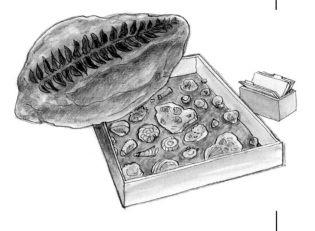

Students were given hand lenses, stereomicroscopes, old dental tools, nails, and other tools to clean and expose more of the fossils as well as glues to fix broken specimens. Some students had worked in other science classes and in summer science programs with the school's recently acquired scanning electron microscope (SEM). Eventually they asked if this instrument might aid their fossil study. In many cases the fossils were too big or lacked enough detail, but some students had specimens with small details that lent themselves to SEM study. An advantage of using the SEM was the instrument's photographic capabilities—images could be preserved and shared with other class members.

The students were given five class periods to work with their specimens and reference materials, to go to the library, and to use the SEM. Partway into their study, students noted that particular fossils seemed to be found with particular rocks. Anna suggested that the location of each fossil be noted on a local geologic map to see what that information would add to the fossil interpretation. When the plotting was done and the map key
continued next page

studied, students were able to make many inferences about the age of particular fossils and the various environments in geological history.

After a week, the groups presented what they had discovered. After just a few oral reports, it became evident that the area had gone through many environmental changes. Anna suggested that the students correlate their studies of fossil environments with what they had learned about plate tectonics and paleography earlier in the year.

The geologic map indicated that they had fossils from the Precambrian, Cambrian, Ordovician, and Pleistocene Ice Age times. Their fossils included cyanobacteria (blue-green algae), an abundance of brachiopods, many types of corals, trilobites, cephalopods, graptolites, mammoths, pelecypods, gastropods, crinoids, bryozoans, and one beautiful starfish. There were some specimens that could not be identified, including some Cambrian fossils that are only now being classified by paleontologists. The Precambrian fossils were lumps of graphite—students found they could write history with a piece of history!

The project motivated students to go out and collect many more specimens, and a volunteer Saturday field trip was made to a few of the good fossil-collecting locations in the area.

The students determined that the fossils from various eras represented many different environments such as ocean troughs, continental shelves, deltas, and the tundra and taiga environments of the Pleistocene Ice Age. These interpretations were then tied to the ocean margin, island arcs, young mountain chains, coastal plains, and glacial environments of eastern United States in the last 600 million years. Much to their surprise, the students discovered that some of the bedrock on the school property was a reef deposit made by cyanobacteria, which may have been the first life form to use photosynthesis and release free oxygen to Earth's atmosphere.

Each group was asked to submit a written report with drawings, photomicrographs, photocopies, or other graphics for a portfolio assessment. The student presentations generated a great deal of discussion, and student preconceptions about the age of the rock formations in their immediate environment were quickly challenged. When Anna's class finally moved from their problem-solving investigations to lessons on what others had discovered about Earth history, they were ready and eager to learn.

After decades of teaching, Anna has never stopped learning. The methods she uses to help students learn in the content-standard areas vary with each class but always involve student-directed inquiry and exploration. Each year her fossil collection and her collection of methods for supporting learning grow.

Origin and Evolution of the Universe

Nature of the Learner

The vast scales of time and distance involved in astronomy provide the most difficult conceptual challenge for beginning students of Earth science. Conceptualizing a time scale more than four billion times as old as they are is almost impossible for most adolescents. Many concrete examples will be necessary if students are to bridge the gap from concrete to formal. However, the intrinsic interest of space science can provide the motivational edge to help students tackle difficult concepts.

History and Nature of Science

The modern study of cosmology is the newest field of geology—and one that has only been fully explored in this century. The fossil photons of the "Big Bang" were detected 30 years ago. Just three years ago, the ripples in background radiation suggested to scientists that the expansion of the universe was not smooth but wrinkled. This data—which may help explain the origin of stars, planets, and Earth—is only just being read by telescopes like Hubble.

Since 1990, astronomers have
- discovered two solar systems around stars other than our own
- photographed the birth of stars in remote galaxies
- confirmed the existence of black holes
- found the birthplace of comets beyond Pluto
- located the first "brown dwarf" star and four new planets.

Nature of Instruction

How do we convey a science that is in its infancy (and may not have existed when many of us were in college)? The Standards suggest that we model fascination, interest, and scientific curiosity and share them with students as we learn together. Reading updates, whether printed or electronic, can help students become cosmologists and can help instructors teach the spirit of inquiry in the context of this content area.

Because much of today's astronomy and cosmology is so new, concrete laboratory exercises may be hard to find. We should feel free to innovate and illustrate. When a news article refers to billion-year time scales, a modeling exercise may be in order. When the mathematics of an astronomical theory seems intimidating, a qualitative example may suffice.

Assessment

In an area as new as today's cosmology, we may want to turn to the Science As Inquiry standards. All students should understand how to identify questions and how to use technology to improve communication. Asking students to develop questions and find sources from which to elicit answers can be a valuable assessment of methods.

Personal and Social Perspectives

Cosmology is one of the best areas for learning about the nature of hypothesis versus theory. The "Big Bang" theory is still in its infancy, and the nature of the international effort to validate or refute it is a valuable lesson about the work of the scientific community. International cooperation (and lack of it), societal acceptance of new ideas, and the influence of radical theories on social standards can all be discussed in the context of astronomy.

Science and Technology

The "golden age of cosmology" has occurred because of rapid increases in technology. The Hubble telescope is perhaps the most widely known of the new instruments. Within the next decade, NASA will put computer-assisted telescopes on the Space Station and in orbit beyond Jupiter. Students can explore the function of the instruments along with the data they produce and in the process relate the resolution that the instruments achieve to the power of the theories they support.

A Classroom in Action

The planetarium in Robert Duggan's school, built in 1965 with National Science Foundation (NSF) funding, is still an exciting place for exploration. By integrating a series of exercises in the planetarium with simulations and investigations, Robert has been able to help high school students construct knowledge about the age of the universe even though they cannot explore the concepts directly.

Months before they begin to study astronomy formally, Robert stimulates students' interest in the field. First, when the class studied latitude and longitude, he arranged a planetarium exercise on sky locations. Second, he invited interested students and their parents to a rooftop sky observation session once a week, during which he provided basic information on telescopes and instructed participants on finding particular objects in the night sky. They observed the Moon, several planets, star clusters, and some galaxies. Third, he arranged for students to sign out and take home telescopes for extended viewing projects. By the time the formal astronomy unit begins in December, enthusiasm is high.

Robert realized that the abstract concepts of the Origin and Evolution of the Universe would be difficult for most secondary learners, so he designed a unit that included lessons for several different learning styles. In a simulation designed as part of Project Spica (coordinated by Darrel Hoff of the Harvard–Smithsonian Observatory), students positioned themselves on the football field representing galaxies in their relative positions in the sky. Each galaxy was assigned a velocity and spectral information as viewed from Earth. Students learned that most galaxies were moving away from the viewer on Earth.

Robert also used a series of pages of dots (representing galaxies). The students were to figure out the proper sequence of the pages based on the spacing of the dots (galaxies). These lessons evolved into a discussion on the origin of the universe and the theory of an expanding universe.

Using texts, astronomy magazines, and NASA resources via e-mail, students gathered data on the ages and location of galaxies and grouped them. They often found the three- and four-dimensional nature of the data difficult to understand, so Robert decided to use the

planetarium again. With flashlights, the students pointed out galaxies at similar distances from the Earth. They inferred that galaxies with similar distances were found at all directions in space! After they explored the concepts of the red and blue shift, they came to understand the Hubble constant.

Students wondered how galaxies behind the center of our own Milky Way could be "seen." Robert presented the students with data he had obtained from the Greenbank, West Virginia, radio telescope, and he allowed them access to more NASA data. As a benefit of attending an NSF workshop at Greenbank, Robert was able to arrange a field trip to the radio observatory. His students earned money for the trip by giving planetarium shows for parents. At the ob-

continued next page

servatory, students used the 40-ft. radio dish for three days and recorded signals of planets, the Sun, and various galaxies. All participants at Greenbank were made to feel a part of real science—they found themselves grappling with many of the questions humans have always been trying to answer.

At Robert's school, moving toward the Standards won't wait on the purchase of significant amounts of new equipment. He already has most of the components of an outstanding program in Earth and space science—enthusiasm, inquiry, and access to technology. Perhaps most important, because there are a variety of ways to achieve in his class, all of Robert's students have an opportunity to achieve high-level outcomes in science.

A Classroom in Action

Integrating the Sciences and More

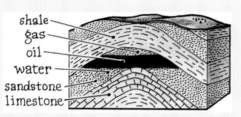

The school board in Gary Ciccetti's district mandated that all high school students take three years of science. Most college-bound students took the full sequence of biology, chemistry, and physics; but other students had few options after physical science and biology.

Gary and his fellow teachers developed an integrated course in Earth and environmental science for grade 11, which presents key Earth and life science content in the context of societal issues. The course offers a computer-assisted review option for the prerequisite physical science information that students might need. Because many third-year students had special needs, a co-teacher specializing in learning disabilities was assigned with Gary for two periods a day.

One of the course's most successful units has been on fossil fuels. The unit begins with a discussion and pretest to determine students' misconceptions. Gary projects cartoons, drawings, and assertions on the board and asks students whether they agree or disagree with the concept presented in each one. He discovers that

- Most students believe there are "lakes" of oil below the ground.
- Almost all believe that oil can be used directly from the well as a fuel.
- About half have an accurate sense of the length of time between the Mesozoic and Cenozoic eras.

On the second day of class, each group finds a pile of "junk" on their lab table—Styrofoam™, plastic forks, ballpoint pens, lead pencils, nails, hand lotion, aspirin, pantyhose, rags, old tests, and soda cans. Gary asks each group to sort the items into those that are made from petroleum and those that aren't. During the discussion, students are surprised to find out that almost all the items on the pile come from petroleum.

On the third day, Gary has groups plot petroleum statistics from the *International Petroleum Encyclopedia* and the *Energy Statistics Sourcebook* on a large map. Many students need extra help to remember the meaning of latitude and longitude. The co-teacher does some oral review, and two computers with review programs are ready for special practice in the resource room. The review of coordinates also helps students remember how to do graphs—a skill they will need later in the unit.

Gary teaches the students that the United States imports half of its petroleum from just three countries: Saudi Arabia, Venezuela, and Mexico. He divides the students into three research groups, one for each country. For each area of the world, the instructors create a list of topics that might be included in the reports. One member of each team is asked to scan news services on the Internet for information about international relations with the country; one member does a text-based report; one prepares a large map; and the fourth establishes a network connection to an international agency in that country. Because students in each group vary in ability and learning style, options for reports and presentations are available.

continued next page

124 Earth and Space Science

The second week of the unit focuses on life science. Students conduct a laboratory in which they determine the amount of oxygen produced by a fresh sprig of *Elodea* in strong and weak light. They review the principles of photosynthesis and look at several macroscopic plants that produce oils (for example, sunflowers and peanuts). They examine photosynthetic protists under the compound microscope, and they use stereomicroscopes to examine fossils trapped in sedimentary rocks. Students with learning disabilities often need extra time in the resource room to review photosynthesis. Those students who demonstrate quickly that they remember their biology participate in computer-based simulation games that explore the relationship between the intensity of light and the amount of carbohydrate produced.

Gary asks groups to prepare slide presentations (using Harvard Graphics or PowerPoint) about how microscopic organisms in the ocean photosynthesize, die, and are converted to hydrocarbons by heat and pressure. Gary gives each group depth and temperature data to plot, as well as bit-mapped diagrams they can incorporate into their presentations. Each group shows its presentation to the whole class for discussion and critique.

Because any unit on oil must be a bit messy, Gary starts one class by asking students to speculate how much motor oil they can pour into a 1,000-mL beaker filled with sand and gravel. When they pour in the oil, they astounded by the amount the beaker can hold. (And the advanced chemistry class is equally astounded when they are given the assignment to clean up the sand and gravel!).

Gary's unit doesn't end with a test; assessments occur every day of the unit. His co-teacher provides assessments of comprehension, reasoning, and language development by moving through the room asking questions of specific students. One way Gary assesses understanding is to ask students to pick a place to drill for oil from a variety of cross-sections showing subsurface geology. He asks students to justify their choice in an essay. A computer simulation of the same task is available for students to use in spare moments; and after the first test, it becomes a popular activity.

Gary's Earth science course continues to develop. Each year new components are added. The American history class has proposed a foreign policy unit based on petroleum resources, and the remedial mathematics class has incorporated some of Gary's graphing exercises. The biology teacher is interested in setting up some ongoing photosynthesis experiments so that students can more easily retain concepts from grade 10 to grade 11. For the teachers at Gary's school, the effort to build new curriculum has also had the benefit of helping them find something in common.

Resources for the Road

American Association of Petroleum Geologists, PO Box 979, Tulsa OK 74101-0979.

National Institute for Petroleum and Energy Research, PO Box 3565, Bartlesville, OK 74005.

National Renewable Energy Laboratory, 1617 Cole Boulevard, Golden, CO 80401.

Oil and Gas Exploration curriculum module. Denver Earth Science Project, Colorado School of Mines, Golden, CO 80401.

Program Standards

To effect change in programs, we wil have to accept positions of leadership in our communities. And to do that, we will have to believe in ourselves.

Moving into the Program Standards

What is your school's science program? A book of courses? A list of outcomes? Flowcharts of teacher activities?

As teachers, we often see school programs in incremental bytes: specific units of content taught in defined spaces for limited periods of time. But for students, learning is less the sum of these discrete experiences and more their total response to the school environment—in other words, their unique path through the courses we offer.

We cannot plot our path toward the Program Standards by listing good courses. They are only the paving stones for that road. Instead, we must identify the destination of each student, and then find as many routes as there are styles of learning. Courses, classrooms, and curricula are all important, but in isolation they can't provide answers to the bigger questions:

- Is the program consistent? Is it paced at a developmentally appropriate rate?
- Does the curriculum connect to other areas of learning and to the students' world?
- Are the necessary mathematical tools embedded in and integrated into the program?
- Is the program supported by the time, space, and equipment that is necessary for student growth at each level?
- Does the program give diverse learners equal footing?
- Does the faculty and school community support and continually renew the program?

Good programs are a synergy of courses, teaching, assessment, and community support in a setting that supports flexibility. The six Program Standards address the following issues:
- consistency in program
- curriculum
- mathematics in science
- resources
- equity and excellence
- schools as communities of learners

Assessing Your Program

Science educators continually call for "authentic assessment"—not only for students but also for programs. Program assessments should define in *measurable* terms what students should achieve as they experience the school's science education program *as a whole*.

Like good classroom assessments, program assessments should be woven into the fabric of the program itself, not artificially imposed from without. We must continually look for indications that students are moving toward the goals set for the program. And we must be careful to weigh these results against the "opportunity to learn" that the system offers its students.

Probably the most common reason school programs fall short is lack of consistency. Standard A asks that every school program have clear goals and high expectations for students. But all too often the goals of school programs change as students move from course to course or level to level.

Today's elementary science often demands informational reading, while middle school science builds skills, and secondary science often stresses memorization. While individual courses may look coherent on paper, taken together they may be very inconsistent from the perspective of the student. This inconsistency lowers program quality. When we make the goals of school courses consistent *across* grade levels, we will find that students will live up to our high expectations.

Many Courses— One Program

While they are in school all students learn; In fact, it's impossible to keep them from learning. The challenge for good teachers and programs is to get students to learn the outcomes *we* have designed for them. That's more difficult.

Students are quick to perceive the rules of our curricular games; and when the rules change too often, students may reject the game of school entirely. Different science courses should present consistent expectations for students, framed in a language that is meaningful to them. Only then will they be able to make steady progress toward mastering the Content Standards.

Across the K–12 grade levels and across science disciplines, inquiry is the thread that ties courses and programs together. The Standards state, "The ability to understand and conduct scientific inquiry is an important goal for students in any school program." From the earliest grades, teachers should give students opportunities to engage in and reflect on natural phenomena through the cognitive processes of inquiry (searching, organizing, originating, and communicating). This is the heart of science.

But believing in its importance does not tell us how to implement it, and not all of us have the same background or preparation in scientific investigation. Inquiry skills represent a content area to be learned in the Standards. In order to maintain a consistent program of inquiry-based science, schools will have to provide curriculum and support for teachers less experienced with inquiry.

Another common tool we can give students in every course is mathematics. Without quantitative understanding, the results of scientific discovery become mere rote facts. The ability to draw on what is learned in mathematics class and apply it in science must be learned. Good science programs should integrate the skills taught to students in math courses. For us this will often mean having to learn a new math language, since the NCTM (National Council of Teachers of Mathematics) Standards for Mathematics have broken new ground. In order to effectively tie science to a school's math program, we will need to learn the language of the NCTM standards as well as the methods of tomorrow's mathematics.

Good programs provide coherent messages not only in method, but also in content and perspective. In many school systems an emphasis on societal implications and real-world applications ties coursework together at every level. When students are asked repeatedly to use science knowledge from different courses as a basis for personal decisionmaking, they see the consistency in their science programs and respond to it.

Inquiry, quantitative skills, applications, and decisionmaking can become the consistent mile markers that guide students along the pathways of science education.

The Price of Change

We cannot just define expectations for students without analyzing the necessary conditions for realistic "opportunity to learn." Class size, schedules, storage, and classroom space are all essential for inquiry. Resources become even more crucial when systems are reaching for equity among students of varied abilities and backgrounds. Mature learners may be able to achieve outcomes with texts and paper that disadvantaged or developmentally delayed students must explore through hands-on experience. Limited resources often exacerbate inequity among students and limit opportunity to learn.

Equal opportunity to learn often requires *unequal* distribution of time and personnel. Girls may require more laboratory time to counter pressures during adolescence; minority students may need role models; and students from language-deprived backgrounds may require more opportunities to express science ideas. Equity can be one of the most difficult pathways for us to negotiate as we move toward implementing the Program Standards.

When political factors threaten to limit opportunities for enhancing equity in school programs, we should remember that *every* student has a unique learning style, and *each* student at some point in the science program will require some "compensatory education." When resources are sufficient and flexible, students can learn to access the help they need when they need it. (*Note:* For more information on equity issues, see Teaching Standard E, page 19; Assessment Standard B, page 49; and System Standard E, page 148.)

Communities of Learners

Although programs belong to schools and their communities, we as teachers are closest to programs and thus often first to bring about change. While the components of a program—teaching, assessment, equipment, space—can be studied in vitro, the structural barriers to better programs are often less easily measured.

Very few college educators, for example, know how long it takes to take attendance, how to cope with fire drills during labs, what influence ADHD students can have on a class, or how to help a student make up a test. Researchers can define the math necessary for science, but only a teacher can measure how much mathematics is forgotten during change of class. The systemic barriers to change—lack of continuity, inadequate physical facilities, poor-quality

textbooks, lack of prerequisite mathematics skills, diverse behaviors and learning styles—are only clearly visible at the classroom level.

To effect change in programs, we will have to accept positions of leadership in our communities. And to do that, we will have to believe in ourselves. As science specialists, we can rely on our inquiry skills (searching, organizing, originating, and communicating) to help us influence our communities toward program reform.

Program reform cannot be legislated from above. So what can the leaders of systems and governmental agencies do to encourage better programs? The Standards suggest two kinds of support: time and teamwork.

By building regular, specified time into the school day and year for program development and assessment, schools can recognize the invaluable contributions of teachers to the challenge of restructuring programs. Authentic program assessment cannot be sporadic or accidental, and it certainly can't be effective if left to quick staff meetings at the end of long days of classroom teaching.

Another way to support program reform is to give teachers opportunities to network with other professionals. The stark isolation of a classroom may be difficult for some administrators to comprehend, but it is the daily reality for most of us. Networking with our professional colleagues (electronically, through professional memberships, and through attendance at professional meetings) gives us valuable and continued input, something we need just to stay current in our field. All these activities enrich both school programs and our students.

Good programs are built of discrete courses integrated with real-world science and supported by the community of learners. The path we select to better programs will not be a straight line, but will take various turns from where we are now to get to the vision of the Standards.

From Here to There and Back Again

As a classroom teacher in an average school district, how do we begin to move our program toward the Standards? It may seem simplistic to say "one step at a time," but many innovations fail because they are too ambitious or try to achieve too much without adequate groundwork and resources.

We might begin by asking ourselves three questions:

Question 1. Where do we want to go? If our system has not defined consistent goals for *every* course across the curriculum, that's where we must begin—and it's not easy. Many systems undertaking K–12 program review fail to achieve consistency because they don't start at the beginning—which in the case of school science is really the end point: the school science the community wants to achieve. The beginning of a school science program is *not* the *course*, but the *student*. What does the community want a graduate of our school's science program to look like?

When schools begin program reform by focusing on the product, the graduate, rather than on individual courses, they often come to very different conclusions. It's far easier to come together as a faculty and abandon cherished courses or units when you've been challenged to fit your piece of the puzzle into a larger picture that is already in focus.

When participants in program review agree on what a graduate should know and be able to do, it's time to chart the contributions of each course or unit to the big picture and write course descriptions. Many schools have found that establishing a common language and format for *every* course in the district is invaluable. For example, each syllabus might begin with a one-page description that defines student outcomes in knowledge, skills, attitudes, and applications. Some districts may wish to add recognition of developmental levels, inquiry skills, research or interdisciplinary connections, or a general description of the course's place in the overall program.

The most common pitfall in any program review is turf protection. Secondary teachers are often accused of being reluctant to abandon cherished units or content, but they aren't alone. Changing programs and courses is hard work and can be intimidating for many of us. There are two essential weapons against that kind of inertia: support from administrators and support from peers. We must know that our work will be appreciated, that it will be meaningful, that it will be supported (with resources and by the community), and that our peers will also be making changes with us.

Question 2. Do support systems exist? Many grandiose proposals fail when support that was promised fails to materialize. A district may rewrite a program with a heavy emphasis on lab skills, but then pass the responsibility for buying the materials to site-based committees with varying commitment to the program or to principals who have different priorities. Perhaps even more devastating is the expectation that a lab-based program can appear like magic in a system with few resources and physical accommodations and with the expectation that teachers will provide their own materials.

That's not to say that schools with limited resources shouldn't write programs that will move their schools toward the Standards. But the keys to such changes are consistency and realism. From a statement of philosophy that commits a district to inquiry-based science, a faculty can draw up a plan that says, "As each essential component of the plan is provided, we will take the next step." Unlike the all-or-nothing plans of the 1960s, many new curricular materials offer incremental approaches to lab-based science.

From this point the third question becomes easy to answer.

Question 3. How are we going to get there? This is where our practical experience as teachers is crucial. First, we should identify the systemic barriers to change because only when necessary changes are clearly defined can they be surmounted.

For almost every school, the pathway to better programs includes training. Professional development should never be seen as an inoculation, a one-shot cure for ineffective programming. Staff training must occur before, during, and after implementing a new program. But at the same time, be aware of the "always a bridesmaid" mentality: Some schools believe that they have never had enough staff development to begin change—and yes, some teachers use the need for more and more training as an excuse to delay making *any* change.

Trying a new program is a little like skydiving—with the support of peers and the school community as the parachute. Some training is certainly needed before we begin, but once we have some confidence, an experimental jump (in the form of a pilot unit or module) is often just what we need to build confidence. Often we never know what we need to know until we've tried a new method in our *own* classroom.

Your Program Begins with You

Unlike teaching or assessment, the path to better programs is one that cannot be walked alone. Within your school and system, the professional staff together must define its program destination—the picture of that ideal graduate—and then chart paths toward that goal that parallel and support one another. The Standards can provide guiding principles, but the journey is one that only we can make in our schools.

About Program Standard A

Consistency in Program

Good programs have clear goals and expectations within a consistent curriculum framework. The curriculum framework, teaching practice, assessment, support systems, and the responsibilities for implementation are coherent.

Great Expectations

Although it's been 70 years since George Homans carried out his psychological studies of workers in a Western Electric factory, most educators can clearly describe his now-famous Hawthorne Effect: People generally live up to the expectations we have of them. This is certainly true of students. They do best in programs when goals and expectations are consistent and communicated clearly. That principle has become the foremost standard of excellence for the school science programs of the future.

Setting clear goals and expectations is much more difficult than it seems. When you define a course by what the teacher does, it is possible to do everything you want to do. But when a course is defined by challenging outcomes for students, the work is never done.

The Standards define a good program as one that is designed in terms of student knowledge, skills, and attitudes. Teaching practices, assessment, and support systems all align with those outcomes.

The focus on expectations or outcomes isn't new. The writings of J. H. Block and Benjamin Bloom on mastery learning come to mind. Pilot programs in Johnson City, New York, and Rochester, Minnesota, in the 1980s gained national prominence by reaffirming that students achieve more when their teachers have high expectations of them.

The most recent call to concentrate on student achievement has come from the nation's largest science and science education professional associations. For example, the two largest science curriculum reform projects, Project 2061 of the American Association for the Advancement of Science (AAAS) and the Scope, Sequence, and Coordination project of the National Science Teachers Association (NSTA), are both

PROGRAM STANDARD A

All elements of the K–12 science program must be consistent with the other *National Science Education Standards* and with one another and developed within and across grade levels to meet a clearly stated set of goals.
- In an effective science program, a set of clear goals and expectations for students must be used to guide the design, implementation, and assessment of all elements of the science program.
- Curriculum frameworks should be used to guide the selection and development of units and courses of study.
- Teaching practices need to be consistent with the goals and curriculum frameworks.
- Assessment policies and practices should be aligned with the goals,

continued next page

> student expectations, and curriculum frameworks.
> - Support systems and formal and informal expectations of teachers must be aligned with the goals, student expectations, and curriculum frameworks.
> - Responsibility needs to be clearly defined for determining, supporting, maintaining, and upgrading all elements of the science program.
>
> Reprinted with permission from the *National Science Education Standards.* © 1996 National Academy of Sciences. Courtesy of the National Academy Press, Washington, D.C.

designed around a framework of knowledge, skills, and beliefs for guiding student achievement.

As we have refocused our programmatic lens to concentrate on desired outcomes (effects) instead of causes, we have changed our view about the level of support that is needed to foster achievement. Experience, environment, and time are all variables we manipulate to achieve the results we want.

We have begun to look for "whatever it takes—as long as it takes" to enable students to meet our high expectations. We have changed course schedules, credit systems, graduation requirements, and grading practices in order to demand higher achievement of our students. And often we've run into strong opposition from our communities in the process.

Concentrating on the result, whether we call it an expectation, a performance standard, or an outcome, can dramatically change classroom psychology. And that is where controversies have arisen. The opponents of outcomes-based education systems (OBE) have made several significant points:

- Individualizing a program is difficult, if not impossible, in most public schools because of large class sizes and growing numbers of special needs students. Without adequate resources, everyone gets shortchanged.
- Where expectations are watered down so that virtually every student can achieve them, more advanced students can become bored or lethargic.
- Trying to help all students in a group achieve the same outcome requires such a high degree of differential instruction that students at the top may be shortchanged.
- Students who discover that they can retake tests and take their time at projects often learn to milk the system for time.

Are these valid reasons to abandon an outcomes-based approach? Or are they symptoms of a good idea that has been poorly implemented? Unfortunately, the debate about OBE has become so emotional in many communities that the real programmatic issues have been forgotten.

The Standards call not only for a focus on outcomes, but also for aligning good teaching, assessment, and systems support to make high student expectations a reality. They ask program de-

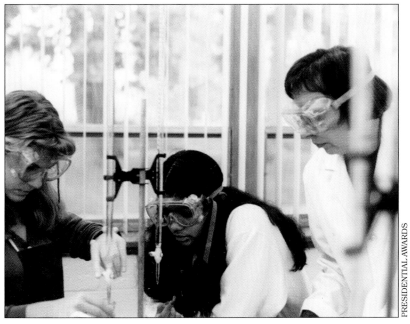

Moving into the Program Standards   133

signers to spend time defining outcomes and to consider how they expect to see (assess) them when they occur. They ask schools to consider what levels of modification and compensatory education will provide optimal "opportunity to learn" for *all* students. Perhaps most important, they ask schools and systems to consider how to support good ideas.

In the factories of the 1920s the workers guessed what the psychologists wanted and made the Hawthorne Effect synonymous with living up to expectations. In the 1990s, our students shouldn't have to guess. By identifying the end points of education, we can debate the issues on a common foundation with *all* stakeholders in education. We can also make our programs more consistent and align them with the expectations of the community and the workplace. High hopes—and high standards—are the bywords for the science classrooms of tomorrow.

Resources for the Road

Kudlas, John M. (1994, May). Implications of OBE. *The Science Teacher, 61* (5), 32–35.

Pratt, Harold. (1995, October). A Look at the Program Standards. *The Science Teacher, 62* (7), 22–27.

Selby, Cecily Cannon. (1993, October). Outcomes-Based Education. *The Science Teacher, 60* (7), 48–51.

Texley, Juliana. (1993, October). Editor's Corner: An Inevitable Outcome. *The Science Teacher, 60* (7), 6.

About Program Standard B

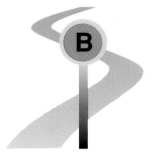

Curriculum

Students discover science content embedded in a variety of experiences that are developmentally appropriate, interesting, and relevant to their lives. Inquiry is a tool for learning science and connecting the science curriculum to other subjects.

The Web We Weave

We cannot find everything we need to know about a good science program in kindergarten—but we can find most of it. There most lessons are built around the letters K-W-L.

"What do you already **K**now?"

"What do you **W**ant to know?"

"Now that you've read, what did you **L**earn?"

That formula summarizes much of modern learning theory. To construct new knowledge, students must first become aware of their own internal world views. In science, we begin by exploring connections to the real world or by conducting demonstrations or activities that challenge students' world views.

We know that science content, as outlined in the Content Standards, can't be taught without context. The Standards suggest that content should be woven through familiar patterns that are "developmentally appropriate, interesting, and relevant to human lives." From mini-lessons that seize the potential of a news event (Wright, 1992) to statewide restructuring programs based on the relationship of science, technology, and society (Yager, 1990), programs that connect to the real world produce measurable gains in student learning.

Integrating science with the real world is often easier than teaching science cold. Summarizing 10 years of restructuring programs around S/T/S, Robert Yager documents that students in such programs "relate their studies...become involved... seek out information...ask more questions...become more curious...see science as skills they can use...retain information and relate it to new situations." Students who learn science in a relevant context achieve many more outcomes simultaneously— and almost effortlessly.

PROGRAM STANDARD B

The program of study in science for all students should be developmentally appropriate, interesting, and relevant to students' lives; emphasize student understanding through inquiry; and be connected with other school subjects.

- The program of study should include all of the Content Standards.
- Science content must be embedded in a variety of curriculum patterns that are developmentally appropriate, interesting, and relevant to students' lives.
- The program of study must emphasize student understanding through inquiry.
- The program of study in science should connect to other school subjects.

Reprinted with permission from the *National Science Education Standards*. © 1996 National Academy of Sciences. Courtesy of the National Academy Press, Washington, D.C.

But we know that programs can't stop with what students already know. The "W" in K-W-L stands for "What you *Want* to know." Good science programs leave students hungry for more and provide inquiry as a tool for continuing growth.

Inquiry has been a hallmark of good programs since the 1960s. Yet most teachers still associate it solely with laboratories. Inquiry isn't an activity, but a state of mind and a way of knowing that can be expressed in many forms: These include questioning, researching, analyzing and synthesizing data, carrying out hands-on investigations, doing surveys, or conducting field explorations. When an inquiring attitude is woven throughout a K–12 program, students move eagerly from one level to the next, always coming back for more.

When inquiry catches on, it's infectious. When any teacher asks a student from a good science program "What have you *Learned?*" the student is quick to apply the scientific habit of mind to other areas. Inquiry-based programs lead naturally to integrated studies because students (unlike faculty) seldom take the divisions among disciplines very seriously. The student who learns how to question, debate, or explore can seldom be stopped.

The lessons of kindergarten are the products of the natural curiosity of children—every young child is already a good scientist. It's only when we forget the wonder of comparing our internal and external universes that science becomes a burden. To move a program toward the Standards, we must weave strong strands of relevance and inquiry through each experience. And as we do, we must tie each class and course to the other courses in the curriculum. The result will be a fabric that students can wear all their lives.

Resources for the Road

Carlson, Patricia. (1993, February). Environmental Investigations. *The Science Teacher, 60* (2), 34–37.

Crandall, Bill, and Varrella, Gary. (1995, October). Issue-Based Science. *The Science Teacher, 62* (7), 42–45.

DeFina, Anthony V. (1995, September). Environmental Awareness. *The Science Teacher, 62* (6), 33–35.

Gowen, Lorraine F., and Marek, Edmund A. (1993, January). Science Fairs: Step by Step. *The Science Teacher, 60* (1), 37–40.

Huebel-Drake, Madeline, Finkel, Liza, Stern, Elizabeth, and Mouradian, Mike. (1995, October). Planning a Course for Success. *The Science Teacher, 62* (7), 18–21.

Pederson, Jon E. (1992, May). Take Issue with Science. *Science Scope, 15* (8), 34–37.

Ramsay, John M., Hungerford, Harold R., and Volk, Trudi L. (1990, March). Analyzing the Issues of STS. *The Science Teacher, 57* (3), 61–63.

Wright, Russell G. (1992, February). Event-Based Science. *The Science Teacher, 59* (2), 22–23.

Yager, Robert E. (1990, March). STS: Thinking Over the Years. *The Science Teacher, 57* (3), 52–55.

Yager, Robert E. (1992, November/December). Appropriate Science for All. *Science Scope, 16* (3), 57–59.

Yager, Robert E. (1993, February). Make a Difference with STS. *The Science Teacher, 60* (2), 45–48.

About Program Standard C

Mathematics in Science

The good science program should be coordinated with the mathematics program to improve student understanding and give students the tools to succeed.

Every Good Program Counts

The revolution in science education can be traced to the dramatic changes in mathematics education in the late 1980s. After March 1989, when the National Council of Teachers of Mathematics (NCTM) released its *Curriculum and Evaluation Standards for School Mathematics*, the third "R" was never the same again.

Although the release of the math standards may have seemed sudden to observers of America's schools, the standards were actually the result of three years of planning, writing, and consensus-building among teachers, scientists, engineers, and the business community. The swiftness of the changes that occurred in that discipline in light of the standards pressured those in other subjects to face the need for change. When we looked across the hall to peers who were emphasizing integration, problem solving, cooperative learning, and hands-on mathematics, we often wondered when our turn would come.

Perhaps the most effective part of the NCTM program was the publication of executive summaries of the standards for administrators and the general public. Stakeholders who had not studied math for many years could read and understand the brief, user-friendly guides. They could ask, "Is our program like this?" and feel some confidence that they would understand the answer.

In many ways, the path to the National Science Education Standards was easier because the math profession led the way. The math teachers built their standards on two major assumptions:
- *Teachers are key figures in changing how mathematics is taught and learned.*
- *Such changes require that teachers have long-term support and adequate resources.*

> **PROGRAM STANDARD C**
>
>
>
> The science program should be coordinated with the mathematics program to enhance student use and understanding of mathematics in the study of science and to improve student understanding of mathematics.
>
> Reprinted with permission from the *National Science Education Standards*. © 1996 National Academy of Sciences. Courtesy of the National Academy Press, Washington, D.C.

Like the science teaching community, the mathematics teaching profession is shifting toward classrooms as learning communities; toward logic, reasoning, conjecturing, inventing, and problem solving; and toward connecting mathematics to other applications. It's the last shift that will help us the most. In many science content areas in the past, we have resorted to direct instruction rather than discovery because students lacked the math skills to construct new ideas themselves.

For example, biology students studying cell size need to understand surface/volume ratios; while in population studies, they need to know about sampling techniques. High school students interpreting the results of experiments need to understand statistical significance. In Earth science, trigonometry is indispensable in mapping, and exponents assist the understanding of radioactive decay.

To integrate math into science, we will have to learn about the NCTM standards and how math is being taught. We'll also have to learn how to tap the math knowledge of our students *outside* their math classes.

Piaget talked about "horizontal decalage" in his research, and many of us know the concept if not the name. A student who functions quite logically in a familiar area shows far less reasoning ability in an unfamiliar context where he or she doesn't expect to have to be logical.

To access prior knowledge and get students to apply what they already know about math to science, we will have to spend time eliciting students' experiences in math class. To do that, we will have to know what is taught as well as the vocabulary used. It will also help if math teachers remind students that they will be expected to carry their skills across the hall!

In many high schools an amazing thing is happening: Teachers in two or more departments are developing thematic activities that push students to apply knowledge across the curriculum. Sophomore geometry students are building cell models (and even writing science fiction stories about giant monster cells!). Freshman Earth science students are solving algebraic equations, and junior English students are writing abstracts of their own research. Learners often come to the inescapable conclusion that every subject counts.

Resources for the Road

Atwater, Mary M. (1995, October). The Cross-Curricular Classroom. *Science Scope, 19* (2), 42–45.

Gilligan, Carol, Lyons, Nona P., and Hammer, Trudy J. (1990). *Making Connections.* Cambridge, MA: Harvard University Press.

Interdisciplinary Approaches in Middle Level Science Education (Theme Issue). (1993, March). *Science Scope, 16* (6).

National Council of Teachers of Mathematics. (1989). *Curriculum and Evaluation Standards for School Mathematics.* Reston, VA: Author.

National Council of Teachers of Mathematics. (1989). *Professional Standards for Teaching Mathematics.* Reston, VA: Author.

National Research Council. (1989). *Everybody Counts: A Report on the Future of Mathematics Education.* Washington, DC: National Academy Press.

Richmond, Gail, and Striley, Joanne. (1994, October). An Integrated Approach. *The Science Teacher, 61* (7), 42–45.

Welch, Wayne. (1978). Science Education in Urbanville: A Case Study. In R. Stake and J. Easley (Eds.), *Case Studies in Science Education.* Urbana, IL: University of Illinois.

About Program Standard D

Resources

Science students must have access to resources, including time, materials and equipment, space, qualified teachers, and the community.

Staging a Great Production

You can find good science programs almost anywhere—but that doesn't mean that *the environment* in which science occurs isn't important. The physical setting for learning greatly influences the direction and quality of science experiences. Communities that want good science programs must support them with physical facilities that accommodate active exploration, positive collaboration, and safe investigation.

The influence of the physical setting on science teaching and learning can't be overestimated. Cronin-Jones (1992) cites a 1988 poll that found that on a given day only about 20 percent of science classes are involved in hands-on activities. Is that because the teachers want to lecture? Or because the challenge of providing materials, flat work surfaces, and group-seating arrangements is insurmountable? Or because the potential levels of sound or chaos in the room are just too daunting? Classes that learn in a small room with desks facing front will find constructing new knowledge difficult.

What should a good science laboratory look like? Surprisingly, the research is slim. Science teachers generally agree that a lab needs the following (for more, see Appendix C: Designing High School Science Facilities):

- enough space to explore (5 square meters per student)
- flat, movable work surfaces (0.5 meter horizontal work surface per student)
- lockable storage
- sinks (at secondary level, 1 per 5 students)
- access to electricity and gas (without using extension cords)
- smoke alarm and fire control equipment
- first aid (including burn care and eye wash)
- ventilation (measured by air movement)
- safe, adequate lighting
- facilities for maintenance

PROGRAM STANDARD D

The K–12 science program must give students access to appropriate and sufficient resources, including quality teachers, time, materials and equipment, adequate and safe space, and the community.

- The most important resource is professional teachers.
- Time is a major resource in a science program.
- Conducting scientific inquiry requires that students have easy, equitable, and frequent opportunities to use a wide range of equipment, materials, supplies, and other resources for experimentation and direct investigation of phenomena.
- Collaborative inquiry requires adequate and safe space.
- Good science programs require access to the world beyond the classroom.

Reprinted with permission from the *National Science Education Standards*. © 1996 National Academy of Sciences. Courtesy of the National Academy Press, Washington, D.C.

Moving into the Program Standards 139

of healthy living things
- separate lockable rooms for teacher preparation
- access to computer(s) for data analysis
- access to telecommunications for research and data gathering

Just as learning styles vary, the set-up of the classroom should be flexible so it can be changed to fit various lessons. Group and individual work should be accommodated, and references should be handy. What is, perhaps, surprising is that such facilities are recommended at *every* grade level—not just the secondary level. In the 1990s students shouldn't have to walk to the library; the CD-ROM tower should be as close as a network, and the "Oz Room" that sends video signals on demand should be accessible.

Including All Students

Another factor that affects the design of science classrooms is inclusion. Most students with physical or learning challenges benefit from heterogeneous class experiences in science. But to welcome these students, many changes in traditional construction have to occur, including
- wider doors
- lower controls, paper towel dispensers, etc.
- heavier furniture (but lighter closings on doors)
- Braille on some equipment
- hook-ups for earphone jacks and other personal computer devices

A common characteristic of good science classrooms—specimen collections and displays—can actually hinder learning for ADHD students. A classroom that is visually busy can be an advantage to students with normal attention spans, but very distracting to others.

Classrooms can be constructed with study carrels in some locations to reduce stimuli during testing and silent reading times. These same carrels can be invaluable to visually impaired or learning disabled students who need taped versions of print material. (Books on tape are readily available—ask your special education department.)

We often feel that we are powerless to affect the physical environment in which we teach because it is costly to remodel. But the most powerful argument of all is laboratory safety. Every secondary classroom, for example, should have a separate, locked storage area with sliding track shelving. Nonflammable chemicals should be

Resources for the Road

Abend, A. C., Bednar, M. J., Froehlinger, V. J., et al. (1979). *Facilities for Special Education Services: A Guide for Planning New and Renovated Schools.* Reston, VA: Council for Exceptional Children.

American Association for the Advancement of Science. (1991). *Laboratories and Classrooms in Science and Engineering.* Washington, DC: Author.

Ball, John. (1996, March). Technology Infusion Strategy. *The Science Teacher, 63* (3), 51–53.

Biehle, James T. (1995, November). Six Science Labs for the 21st Century. *School Planning and Management, 34* (9), 39–42.

Charron, Elisabeth, and Woolbaugh, Walter. (1994, November). A Room with a View. *The Science Teacher, 61* (8), 38–41.

Clemmitt, Sarah. (1996, March). Accessible Internet Data. *The Science Teacher, 63* (3), 48–50.

Cronin-Jones, Linda L. (1990, March). A Lab for All Reasons. *The Science Teacher, 57* (3), 36–41.

Cronin-Jones, Linda L. (1992, October). Is Your School a Dumping Ground? *The Science Teacher, 59* (7), 26–31.

Dagher, Zoubeida R. (1995, September). Materials Speak Louder Than Words. *Science Scope, 19* (1), 48–50.

Dougan, David. (1994, November). Hidden Hazards. *The Science Teacher, 61* (8), 25–29.

continued next page

stored on chemical resistant shelving, acids in a special cabinet beneath a fume hood, and flammable chemicals in a specially designed metal cabinet. Refrigerators and lights should be spark-free. Exhaust fans are a must, and every room should have phone connection to the office and to outside emergency help.

We not only have the responsibility to inform ourselves and our students about the potential hazards of exploratory science, but we also have the obligation to inform administrators (often and in writing) about potential safety hazards in the school. Plugs without ground fault interrupters, gas jets without safety cut-offs, and cabinets without locks can all be fixed. Reconfiguring and adding learning space may have to wait until the next round of construction in your district, but they should not be ignored.

Space is only one dimension that defines great science education. Time and training (covered in other sections of this book) are vital parameters, too, as is the institutional structure through which we obtain materials on a regular basis. Class size is one of the most easily manipulated variables of the physical environment. It affects what is taught and what is learned. Administrators who want more lab science will have to consider this variable.

Each of these physical factors alone can't make learning occur, but lack of support in any one area can become a barrier to progress on your school's journey toward the Standards.

Egelston-Dodd, Judy (Ed.). (1994). *A Futures Agenda: Proceedings of a Working Conference on Science for Persons with Disabilities.* Cedar Falls, IA: University of Northern Iowa.

Frinks, Ronald M., and McNamara, David B. (1985, March/April). The Wheelchair-Bound Student in the Physics Laboratory. *Journal of College Science Teaching, 14* (4), 416–420.

Holliday, William G. (1992, January). Should We Reduce Class Size? What the Research Really Says. *The Science Teacher, 59* (1), 14–17.

Mackin, Joan, and Williams, Florence. (1995, December). Science in *Any* Classroom: How To Cope with Less-Than-Adequate Facilities. *The Science Teacher, 62* (9), 44–46.

Madrazo, Gerry M., Jr., and Motz, LaMoine L. (1993). *Sourcebook for Science Supervisors* (4th ed.). Arlington, VA: National Science Teachers Association (NSTA).

Melchert, Sandra A. (1996, February). Bidding Basics for Stretching School Science Dollars. *Science Scope, 19* (5), 34–36.

North Carolina Museum of Life and Science. (1990). Sharing Science with Children: A Survival Guide for Scientists and Engineers. Durham, NC: Author.

North Carolina Museum of Life and Science. (1992). Sharing Science with Children: A Survival Guide for Teachers. Durham, NC: Author.

Roberts, Renee, and Bazler, Judith A. (1993, January). Adapting for Disabilities. *The Science Teacher, 60* (1), 22–25.

Safety Supplement. (1989, November/December). *Science Scope, 13* (3), S1–S32.

Showalter, Victor M. (1982). *Conditions for Good Science Teaching.* Arlington, VA: National Science Teachers Association (NSTA).

Vos, Robert, and Pell, Sarah W. J. (1990, December). Limiting Lab Liability. *The Science Teacher, 57* (9), 34–38.

West, Sandra S. (1991, September). Lab Safety. *The Science Teacher, 58* (6), 45–51.

What Would a New Science Center Look Like?

In 1992, the teachers in the Berkeley district were excited when the community approved a new science center for Berkeley High School. An active group of stakeholders worked closely with an architect and a planning consultant to determine what an ideal high school science facility should be. The sessions included representatives from administration, the entire science department, custodians and maintenance personnel, as well as the professional planners.

Their initial meetings produced the following criteria for the facility:
- The science facility should be centrally located to give it visibility and to foster interaction and model inquiry.
- Labs should be generic and connected to one another—no "dry" labs for biology or "wet" labs for chemistry.
- Lab design should be flexible enough to accommodate changing curricula.
- Technology should be integrated to support group, lab, and individual activities.
- The science center should not only be a fun place to team teach and learn, but also a place that fosters integration with other subjects.
- A space for large-group instruction should be provided, especially for 9th and 10th graders. Lecture and lab facilities should be separate.
- Teachers should have private space apart from teaching areas. The labs should not be "owned" by teachers, but the offices and storage spaces should be.
- Spaces for storing materials should be consolidated for efficiency.
- Traditional and high-tech areas should be blended.
- Equipment that supports and reinforces science learning should be provided.

Four years later, in 1996, a community near Berkeley approved a similar project for the new junior high school. How will their planning be different in the light of the Standards?
- Labs have a far broader purpose today. Separate areas for reading and research will be available in the new science center with access to CD-ROM "towers" and to telecommunications.
- Lecture areas will all but disappear in this facility. Lecture-like classes will be so rare that one room in the entire wing of the science center will suffice.
- Safety facilities will be identical for life, Earth, and physical science labs. Gas hoods, specialized storage and disposal equipment, accessibility, and fire protection will all be standard in every science area.
- Storage will be greatly expanded as individual student research projects become more common.
- Communication among teachers will be enhanced by placing offices and work spaces in close proximity.

Like Berkeley, the neighboring district will employ the services of a planning professional to make the vision of teachers, staff, and students become reality. But unlike the Berkeley facility, the new science center will be more attuned to individual student research and less to direct instruction.

(*Note:* For more information on setting up a science facility, see Appendix C.)

About Program Standard E

Equity and Excellence

All students in a good science program have equitable access to the opportunity to achieve the Standards.

Leveling the Playing Field

Every child is born a scientist. Children have the nonstop curiosity that prompts them to continually compare their internal world with the input of their senses and struggle to make sense of it all. But sometime between the early school years and adolescence, exploration and challenge lose their fascination, especially for girls. By age 13, girls lag six points behind boys on the National Assessment of Educational Progress test in science (277.3 to 221.3). Only 33 percent of all adult scientists are women.

A nation seeking to compete in the world marketplace hasn't one scientist to waste. Every school program must ensure equity for every child—regardless of gender, ethnic background, or learning style. Other parts of this publication have looked at factors in teaching and assessment that can bias opportunity to learn for children from minority cultures.

The pervasive evidence of gender bias in math and science suggests that almost *every* program has some intrinsic bias favoring the learning styles most commonly found in males. Why?

Researchers have grappled with the problem for at least two decades. One of the simplest explanations is expectations—a sort of gender-based Hawthorne Effect in which parents and teachers *expect* males more than females to like and excel in science. This expectation is apparent in classroom observations of questioning, wait-time, assignment of cooperative roles and laboratory tasks, and by textbook examples and visuals. Well-intentioned teachers are often totally unaware of the hidden bias in their techniques.

But the simple answer—of creating equity—may not be the only one. The pervasiveness of the phenomenon (despite many efforts to increase teacher awareness)

PROGRAM STANDARD E

All students in the K–12 science program must have equitable access to opportunities to achieve the *National Science Education Standards.*

Reprinted with permission from the *National Science Education Standards.* © 1996 National Academy of Sciences. Courtesy of the National Academy Press, Washington, D.C.

suggests that other factors in our programs may work against girls. Another answer may lie in the rate of maturation of girls versus boys and its relationship to the standard sequence of science in schools.

Try as we may, most elementary programs are only minimally hands-on. For the majority of students the first lab-based science programs occur in the middle or junior high years. The boys are still children, fooling around without fear. They shout out answers, try almost anything, and seldom show any reluctance to guess. And where are the girls? Looking around with adolescent self-consciousness to see who's looking at them, wondering if the boys will ever settle down. For the majority of girls, hands-on science comes too late.

There are many things that we can do to create greater opportunity to learn for girls in science. Here are a few suggestions adapted from Pollina (1995):
- Connect science to other subjects and the real world (Program Standards B and C).
- Choose metaphors, examples, and visuals carefully to avoid stereotypes.
- Foster true collaboration. (Don't let the boys take over!)
- Give girls the tools and the technology (even if you have to create after-hours opportunities for the science-shy).
- Provide opportunities for "play." Girls may not have been given the opportunity to use hammers, screwdrivers, technology, and other tools.
- Convey a value for verbal strengths (areas in which girls may excel).
- Vary assessment methods (to reward varied learning styles).
- Keep expectations high for everyone.

Even if every science textbook had appropriate mixed-gender examples and every elementary classroom were lab-based and every teacher were bias-free, some embedded bias would still exist in science programs because girls *do* learn differently than boys at some ages.

Pollina suggests that we have focused on girls as if they were a problem, that is, how do we change girls so more of them take math and science? Should we be satisfied in making girls more aggressive, more analytical, more competitive? Or is their something askew with the bias of the methods of science itself, which undervalues what are often considered female approaches to problem solving? Barbara McClintock attributed her Nobel Prize to her "feeling for the organism"; Dian Fossey developed new methods in ethology *because* of her unique perspective. Marie Curie, Anna Freud, and Carolyn Herschel were not echoes of their noted male relatives, but recognized scientists in their own right.

Like differences in learning style, ethnic background, and cultural perspective, differences in gender must be valued in the science classroom. As we accommodate a range of development in each class, we must also accommodate the range of perspectives in our programs. That means broader opportunities to learn for *every* student.

Resources for the Road

Carey, Shelley Johnson (Ed.). (1993). *Science for All Cultures.* Arlington, VA: National Science Teachers Association (NSTA).

Kahle, Jane Butler. (1985). *Women in Science.* Philadelphia: Falmer.

Peltz, William H. (1990, December). Can Girls + Science - Stereotypes = Success? *The Science Teacher,* 57 (9), 44–49.

Pollina, Ann. (1995, September). Gender Balance: Lessons from Girls in Science and Mathematics. *Educational Leadership,* 53 (1), 30–33.

About Program Standard F

Schools as Communities of Learners

Schools must be communities that encourage, support, and sustain teachers as they implement effective science programs.

Empowering Teachers To Teach

As teachers, we see the systemic barriers to change up close, and we hold the keys to making changes.

But systems are often structured so that participation in administrative change is difficult. We may see our relationship to administration as antagonistic. We may avoid confrontational positions because we fear reprisal. But most often, we don't get involved in administrative decisionmaking because of the most common barrier of all—fatigue.

Good instruction takes everything that most professionals have to give. We stretch to support our students and our peers. Short faculty meetings at the end of long days don't provide a forum for real consensus-building or the development of common vision; and release time means preparing for substitutes and grading even more papers than usual. By default, many of the major decisions of schools and systems are made by people who have been out of the classroom for many years.

Program Standard F describes a very different vision: a school as a community of learners that includes students, teachers, parents, administrators, and concerned supporters from around the community. This membership in a network of reform extends from the classroom, throughout the community, and beyond.

How do schools structure change to focus on teachers as the primary agents of that change? The first key is time. It's easy to set up the major questions of change as yes/no votes; it's much more difficult to develop consensus. Voting creates winners and losers. Consensus-building involves examining an issue as many times as it takes to come up with a solution that every stakeholder can live with and ultimately support, even if it would not be each contributor's ideal solution.

PROGRAM STANDARD F

Schools must work as communities that encourage, support, and sustain teachers as they implement an effective science program.

- Schools must explicitly support reform efforts in an atmosphere of openness and trust that encourages collegiality.
- Regular time needs to be provided and teachers encouraged to discuss, reflect, and conduct research around science education reform.
- Teachers must be supported in creating and being members of networks of reform.
- An effective leadership structure that includes teachers must be in place.

Reprinted with permission from the *National Science Education Standards*. © 1996 National Academy of Sciences. Courtesy of the National Academy Press, Washington, D.C.

Consensus-building begins when we as teachers join forces with our communities to create a comprehensive definition of the "product" that we want to create *together:* the scientifically-literate graduate. The Standards begin to suggest a definition:

> "Scientific literacy is the knowledge and understanding of scientific concepts and processes required for participation in civic and cultural affairs, economic productivity, and personal decisionmaking."

But the implications of that definition will vary in each community. The development of consensus on "graduation outcomes" or "exit expectations" of students in a school system is a valuable process, and we should take the lead in it.

Bybee (1995) suggests that the definition of scientific literacy changes each decade. Today's programs emphasize science, technology, and society interactions and multidimensional outcomes. He points out that while it may take a year for groups to agree on statements of purpose, the path from purpose through policy to real change in program and practice can take as long as 10 years.

But while national groups focus on purpose and committees of stakeholders coordinate adoptions (outcomes, scope and sequence, texts), only *teachers* change practice. Without our involvement at the outset, change is impossible.

Becoming Decisionmakers

Harold Pratt (1995) goes further, calling for a change in the culture of the school so that teachers are included in every decision about program development. That goes far beyond selecting a textbook in a once-every-five-year cycle and beyond voting on the speaker for the next inservice day—it involves real decisionmaking.

To make such involvement a reality, many schools will have to consider new hybrid positions in which teachers spend some time each day in leadership positions and some time with students. (That may force major reconsideration of union contracts and pay structures.) But the result will be dynamic involvement rather than nominal buy-in.

Program Standard F maps an idealistic destination, but one that might be reached by taking very practical first steps:
- Develop permanent groups for consensus-building that incorporate teachers as well as all other stakeholders in the school community.
- Schedule regular time for these groups to meet whether or not there is a deadline or a textbook adoption at stake.
- Support teachers in professional networking through publications, memberships, electronic access, meetings, and inservice programs.
- Explore hybrid positions in which teachers spend time in administration while continuing to teach.
- Create avenues of communication with and among teachers to discuss the systemic factors that influence the classroom.

We are not only employees of schools, but like students, parents, and administrators, we are also learners. Systems that want good programs make it easy for teachers to build them.

Resources for the Road

Andersen, Hans O. (1994, September). Teaching Toward 2000. *The Science Teacher,* 61 (6), 49–53.

Brandwein, Paul F., and Glass, Lynn W. (1991, March, April, and May). A Permanent Agenda for Science Teachers: Parts I–III. *The Science Teacher,* 58 (3), 42–46; (4), 36–39; (5), 22–25.

Bybee, Rodger W. (1995, October). Achieving Scientific Literacy. *The Science Teacher,* 62 (7), 28–33.

continued next page

Bybee, Rodger W., and Champagne, Audrey B. (1995, January). The National Science Education Standards. *The Science Teacher,* 62 (1), 40–45.

National Science Teachers Association. (1997) Decisions Based on Science, Arlington, Va: Author.

Jacobson, Willard J., and Lind, Karen K. (1992, March). Progress in Science Education: How Can We Achieve It? *The Science Teacher,* 59 (3), 38–40.

Pratt, Harold. (1995, October). A Look at the Program Standards. *The Science Teacher,* 62 (7), 22–27.

Risner, Gregory P., Skeel, Dorothy J., and Nicholson, Janice I. (1992, September). A Closer Look at Textbooks. *Science and Children,* 30 (1), 42–45, 73.

Changing Emphases

The National Science Education Standards envision change throughout the system. The Program Standards encompass the following changes in emphases:

LESS EMPHASIS ON	MORE EMPHASIS ON
Developing science programs at different grade levels independently of one another	Coordinating the development of the K–12 science program across grade levels
Using assessments unrelated to curriculum and teaching	Aligning curriculum, teaching, and assessment
Maintaining current resource allocations for books	Allocating resources necessary for hands-on inquiry teaching aligned with the Standards
Textbook- and lecture-driven curriculum	Curriculum that supports the Standards and includes a variety of components, such as laboratories emphasizing inquiry and field trips
Broad coverage of unconnected factual information	Curriculum that includes natural phenomena and science-related social issues that students encounter in everyday life
Treating science as a subject isolated from other school subjects	Connecting science to other school subjects, such as mathematics and social studies
Science learning opportunities that favor one group of students	Providing challenging opportunities for all students to learn science
Limiting hiring decisions to the administration	Involving successful teachers of science in the hiring process
Maintaining the isolation of teachers	Treating teachers as professionals whose work requires opportunities for continual learning and networking
Supporting competition	Promoting collegiality among teachers as a team to improve the school
Teachers as followers	Teachers as decisionmakers

Reprinted with permission from the *National Science Education Standards.* © 1996 National Academy of Sciences. Courtesy of the National Academy Press, Washington, D.C.

System Standards

All parts of the education system must work together and in the process support us in moving toward the vision of the Standards.

Navigating the System Standards

We work magic every day. But it is too much to expect that we alone will move the mountainous education system toward the Standards. The Standards are adamant on this issue:

> "[I]t would be a massive injustice and complete misunderstanding of the Standards if science teachers were left with the full responsibility for implementation. All of the science education community—curriculum developers, superintendents, supervisors, policymakers, assessment specialists, scientists, teacher educators—must act to make the vision of these Standards a reality."

All parts of the education system must work *together* and in the process *support* us in moving toward the vision of the Standards. The Standards themselves provide a common language with which teachers, legislators, principals, school boards, administrators, parents, business persons, and industry leaders can dialogue about the direction science education will take.

In the natural world the term "system" is synonymous with coordinated effort and common purpose. But outside the classroom door, the enormity of the education system is often a reason for us to throw up our hands in frustration—so many participants and so few connections between them.

Dissecting the System

Let's look at the system that energizes school science. Most directly, the power structure of most schools can be traced through department heads to the school and district administration and school boards. But despite the common wisdom of local control in America's schools, much of what is decided in K–12 education is outside the authority of these local administrators.

State education agencies implement federal and state legislative mandates, develop outcomes, and draft achievement tests. Alumni of colleges and universities become voters and parents who judge schools based on the successes and failures of their own K–12 education. Business and industry leaders define the qualities they want in workers and rate employees on their K–12 schooling. Parents want input into their children's education, senior citizens want low taxes—and each stakeholder creates resistance to reform.

In addition, the lack of a consistent national education policy coupled with the complexity of a system of 15,000 autonomous school districts are major obstacles to change in education.

The messages we receive as teachers from various stakeholders are contradictory: Is our success judged by students' scores on the state tests? The strength of the advanced placement program? The number of potential college science majors? The total enrollment in science electives in 11th and 12th grades? Student or parent satisfaction? Or by the nation's competitiveness in the world marketplace or its position on international student assessments? With so many judges holding different yardsticks, is it any wonder that many of us feel frustrated?

Now, however, the Standards offer a new voice and a clear direction toward systemic change that supports the work we do as teachers.

Moving Toward a Mission

The seven System Standards address the following issues:
- common vision: the foundation for change
- coordination: speaking with one voice
- continuity: giving things time
- resources: an indispensable part of policy
- equity: saying science for all and meaning it
- unanticipated effects: unintended effects on classrooms
- individual responsibility: we are all responsible for change

The first step toward change in science education is setting the mission. Formulating consistent policy at *every* level is critical. In this process *no* stakeholder can be left out. Parents, community members, business and industry leaders, legislators, and representatives of interest groups are *all* important.

As we bring these other stakeholders into the policy dialog, we must be vigilant to avoid an inherent source of conflict: When it comes to school policy, everyone believes they are experts because they have all been to school. Also, when the system invites in and recognizes other "experts" in the field, we often feel that the many years we spent studying how learning occurs in our classrooms is not appreciated. When non-teaching "experts" have tried to take over the jobs we do, we know that the results have often been abysmal.

In the dialog on working to achieve the vision of the Standards, teachers and administrators must agree on this: We as teachers may be the experts in *how* to move toward educational goals, but we *never* set the goals *alone*.

Seeing Through Others' Eyes

Our perspective from inside the classroom can sometimes be so absorbing that we fail to look at what we do through the eyes of other stakeholders in the education system. Yet knowing the priorities of others can be invaluable in showing us how to influence the system as a whole. Let's sample several viewpoints:

Naomi Smith, Science Education Curriculum Consultant, State Department of Education

After many years working with children, I opted for a two-year contract at the state department of education. I thought that here I could really make a difference. But I find that while much of what I do has broad application, the parameters within which I must operate are much narrower than in the schools in which I taught.

Each year the legislature

passes a school aid bill and often issues modifications to the school code, which mandate changes in what we do. While the bills are being written, our schedule is frantic. We are called to consult on the implications of the legal language for schools. We attend meetings nonstop with superintendents, lobbyists, and legislators. Not everything that appears in the final document is subject to such scrutiny, because special-interest groups often introduce language at the 11th hour that ends up in the compromise document, and we are left to figure out how to implement it. And on top of all that, there's no guarantee that the changes the legislature demands will be funded.

Perhaps our most rewarding role is when we can act as consultants to schools and systems as they move toward national mandates. Teachers often invite us to help convey "best practice" in their district, in the belief that our position will help them influence local experts.

In this sense, the Standards are a great help to state agency personnel. We will have the authority of a common voice, and our efforts to provide direction to the legislature and school boards will be more focused.

Melvin Jones, Facilities Planner for a Multinational Technology Manufacturer

Science education may be a set of standards to you, but to me it's the foundation of profitability. I need workers in my plants who can read directions, operate advanced robotic machinery, and think on their feet. Show me the best school district, and I'll put my plant there.

Don't take this personally—but our company has been very frustrated with educators. We invite them in to help us with the problem of workforce training, and all we get is philosophy. We ask them where the school science programs are the best, and they cite test scores. When you have to shut down a line because the workers can't keep up with the equipment, talk is cheap.

We are firmly committed to helping science and math education efforts. We are ready and willing to contribute cash and personnel support to a school system partner. But we need that school to look at the end product of their programs from our perspective. Training kids for more book work in college just won't make it.

Show me a school system that is training kids to think and to communicate, and we'll provide the hands-on equipment and mentor-partners to make their programs realistic.

Business isn't the enemy; we're just an undervalued partner.

Anna Rodriguez, School Superintendent

Most days I feel a bit schizophrenic. In my heart I'm an educator, but in my head I am the CEO of a $25 million corporation—the biggest employer in town. There's no way to describe what we do on a given day because the main job of Central Administration is juggling—school board members, taxpayers, parents, students, teachers, noninstructional employees, unions, insurance agents, contractors, and regulatory agencies.

The purpose of Central is to support schools and teachers. But sometimes the help we provide seems like a bitter pill to the individual classroom teacher. It's hard to make teachers see that keeping this business solvent and consistently implementing contracts are vital parts of the support that we provide, and that the long-term benefits of efficiency are worth the short-term frustration of rules.

The greatest philosophical barrier we run into is site-based management. School staffs are the experts in finding solutions to many problems. But often the solutions they propose go far beyond their own sites and impact other cost centers. Then we have to set limits. Here are some examples:
- School A sets a schedule that delays the buses, so school B's schedule is set back.
- School X has so little storage space that the elementary teachers do very little hands-on science. When their students get to the junior

high, they do poorly in the advanced classes.

- *Our enrollment is decreasing at the high school, and the contract gives veteran teachers the right to bid into the middle school to teach science even if they don't have any training in science at all.*
- *The high school science teachers want to leave experiments running all night. The custodians have filed a grievance.*

The most significant support I give individual teachers is to help them cut through the paperwork jungle. Teachers need to help us identify the frustrating parts of our administrative guidelines. We do try to listen to input. There are many times that we look at a teacher complaint and say, "Do we do what we do because it's always been done this way? Or did we try it differently once and it didn't work?"

But teachers, in turn, need to help us find little ways to save money. The shipping charges on six small orders can be vastly reduced by sending one big order instead. Opening a window may be easier than calling the custodian to turn down the heat, but it isn't cheaper.

Central's not the enemy. Yes, we have to think like a business—but we're a nonprofit business, and every saving we achieve can help make someone else's dream a reality.

George Stone, School Board Trustee

I didn't run for the Board to push through my own personal agenda; I ran to make the school better for our community's children. I think I can contribute a different perspective. My children, who have graduated, were dissatisfied with this district, but I hope that children who are in the schools now won't be.

I have an ear to the community, in the coffee shops and grocery stores. I can't go anywhere without someone telling me about the schools. It's usually a complaint because human nature is such that people don't often volunteer compliments. I try to remain at a policy level, but it's hard sometimes. When you get calls at home, you have to follow up on them. Then I go to the school and talk to the teacher to find out what's going on.

I want my schools to shine. I want parents and taxpayers to be happy. I certainly don't want to have to vote on cutbacks and student discipline problems. That's not what why I ran for the Board.

Making the Invisible Visible

In many states the National Science Foundation has funded Statewide Systemic Initiatives and Urban Systemic Initiatives that predate the Standards by three to five years. While these projects take different forms in different jurisdictions, their purpose is to examine the varied components of the educational system with an eye toward bringing all the stakeholders together to commit themselves to work toward change. They begin by asking questions: What kinds of changes succeed? Why do many efforts fail? Why do some begin with vigor and then die?

What seems simple on paper often becomes extremely complicated to put into practice. This can be true whenever groups of any sort get together to work toward change. Consider this scenario:

Averagetown's School Board wants better science programs. After studying the successes of a program in a neighboring state that presents concepts in an "applied" context, the Board mandates that the program be implemented in Averagetown. The Board appropriates funds for inservice training.

Average High has eight science teachers. Two volunteer for the training, which takes them out of school for three days in May. They write the necessary requisitions, and the equipment for the new program appears in their classrooms in August. A flyer is developed and a video shown on the intraschool channel. Two hours of the new course are scheduled.

The program starts off relatively well. The text material is good, but the equipment doesn't always work. Teachers spend many after-school hours phoning the manufacturer and delving into manuals. About half the labs are successful, and another quarter have potential. For the remainder, either the equipment never worked or some was missing. (Adult ed shares the classroom at night, and the meters and gauges might have proved tempting for their activities.) All in all, however, the program seems on the road to implementation.

In March, the students take the state science assessment. Teachers in traditional classes spend several weeks reviewing for that test. Those in the new sections spend minimal time because they are behind in labs. The test scores of the students in the new sections are lower than those of students in the traditional class.

Counselors are busy scheduling courses for next year. When students ask about the new class, the counselors shrug. There are no deliberately negative messages, but they wonder out loud, "Will colleges accept it?" Meanwhile, one of the teachers who volunteered for the program decides to retire.

Enrollment drops. One section is scheduled for year two. The remaining teacher finds that there is no budget provision for replenishing missing or consumable equipment in the program because the Board's attention has shifted to math for the coming year.

...And so goes another good idea following the path of the dinosaurs.

If we want to study the systemic barriers to good science, we must be sure to look in the corners that are often ignored, including counseling approaches, storage systems, custodial resistance, requisition systems, changing priorities, provisions for reassignment in teacher contracts, fickle enthusiasms of evaluators, and more. It doesn't matter how big or small the problems are on the surface; each can be the straw that breaks the back of reform.

There are many forks in the road to better science education. Without *all* parts of the education system working *together*, change can be only minimal at best. We as teachers *cannot* do it alone. Every stakeholder has a role in realizing the vision of the Standards.

In the following pages, let's examine the prerequisites for changing the education system to support science education.

About System Standard A

Common Vision

Policymakers who influence science education must have a vision that is consistent with the vision of those who coordinate teaching, assessment, professional development, and programs.

Why Teach Science?

The revolution of the 1980s in science education began not in classrooms but in the conference rooms of major businesses and the secured meeting rooms of the Pentagon. America's industries were not competing successfully overseas. Military recruits couldn't run the high-tech controls of the new weaponry. Factories were retooling, but workers couldn't keep up.

Teachers who had just learned to accept "back to basics" were sent reeling. As Willard Daggett (professor and advisor to President Clinton) summarized, "We were doing a great job of teaching what was needed in the 1950s and 1960s...." Just when we began to win the old game, the rules changed.

The essence of the reforms of the last decade was the realization that different stakeholders have very different yardsticks with which they measure success in science education. We thought we were preparing students for college; society wanted employable high school graduates. At best, the disparate voices haven't joined in any harmonious dialog; at worst, they have contradicted one another.

A major study of the bipartisan Commission on the Skills of the American Workforce found that 90 percent of American employers did not consider low skill levels in employees a problem and concluded that we still rely on "an outdated system in which a small educated and highly trained elite directs the activity of another three-quarters of our workers with minimal skills" (Fiske, 1991).

But while many experts have urged direct involvement of business in reform, the results of that involvement have been mixed. The Louisiana Association of Business and Industry, after spending eight years in a partnership with one of the lowest-ranked state education systems, reported

SYSTEM STANDARD A

Policies that influence the practice of science education must be congruent with the program, teaching, professional development, assessment, and content standards while allowing for adaptation to local circumstances.

Reprinted with permission from the National Science Education Standards. © 1996 National Academy of Sciences. Courtesy of the National Academy Press, Washington, D.C.

discouraging results: "...[M]ost reforms had been watered down, ignored, not implemented properly, taken to court by teachers unions and others, mired in turf battles, or not funded." From the standpoint of that and other business partnerships, the state educational system was a "gigantic, ever-growing, bureaucratic sponge...."

To move toward the Standards, we must enter into higher-quality dialogs with business and industry. We might begin with the definition of scientific literacy that the *Standards* present:

> "Scientific literacy is the knowledge and understanding of scientific concepts and processes which is required for participation in civic and cultural activities, economic productivity, and personal decision-making."

Once all participants agree on the purpose of their efforts, the ways in which teachers and business and industry can cooperate in the reform effort become limitless. Here are just a few examples from journals:

- The United States Chamber of Commerce and National Alliance of Business have established centers focusing on education and training issues (Usdan, 1992).
- The *New Explorers* series links corporate scientists with students and teachers in grades 7–12 (Moryan, 1994).
- Genentech's Access Excellence online forum at http://www.gene.com:80/ae/ links biology teachers with researchers.
- Utility companies have funds for staff training, libraries, and presentations (Sprague and White, 1992).
- The High Performance Schools Project of JCPenney and the National Alliance of Business have charted major success in developing Project C3 (emphasizing applied learning) (Donnelly, 1992).
- Independent partnerships provide authentic field work in industries across the country (Mackin, 1994).
- Military bases, especially Army bases, can offer mentoring experiences for their communities (Olson, 1993).

The list could go on and on. Once you speak the same language as the business or industry down the road, the revolution in *your* school system will begin.

Resources for the Road

Andersen, Hans O. (1994, September). Teaching Toward 2000. *The Science Teacher, 61* (6), 49–53.

Brandwein, Paul F., and Glass, Lynn W. (1991, March, April, and May). A Permanent Agenda for Science Teachers. Parts I–III. *The Science Teacher, 58* (3), 42–46; (4), 36–39; (5), 22–25.

Bybee, Rodger W. (1995, October). Achieving Scientific Literacy. *The Science Teacher, 62* (7), 28–33.

Bybee, Rodger W., and Champagne, Audrey B. (1995, January). The National Science Education Standards. *The Science Teacher, 62* (1), 40–45.

Donnelly, Melinda. (1992, Fall). Learning to Earn. *America's Agenda: Schools for the 21st Century, 2* (3), 34–41.

Fiske, Edward B. (1991, Spring). Only China-Breakers Need Apply. *America's Agenda: Schools for the 21st Century, 1* (1), 20–24.

Jacobson, Willard J., and Lind, Karen K. (1992, March). Progress in Science Education: How Can We Achieve It? *The Science Teacher, 59* (3), 38–40.

Mackin, Joan. (1994, November). Practical Partnerships. *The Science Teacher, 61* (8), 47–48.

Moryan, James. (1994, May). The Corporate Connection. *Science and Children, 31* (8), 18–19, 38.

Olson, Jean. (1993, February). Military Mentoring. *The Science Teacher, 60* (2), 42–44.

Sprague, Jim, and White, Janet. (1992, November/December). The Utility Connection. *Science and Children, 30* (3), 16–18.

Sussman, Art. (1993). *Science Education Partnerships.* San Francisco: University of California.

Usdan, Michael D. (1992, Fall). Down to Business. *America's Agenda: Schools for the 21st Century, 2* (3), 14.

About System Standard B

Coordination

The policies that influence science education should be coordinated across agencies, institutions, and organizations.

The View from the Top

Parallax is easy to demonstrate: It's as close as a finger in front of your nose, or the administrative office just down the hall from your classroom. The view from different seats in the education system can be significantly different, even when professionals equally devoted to schools and children sit in these seats.

The United States is one of the few industrialized nations without national standards for all school subjects. Local control of schools is a time-honored principle in our nation. But the lack of strong standards has left local boards and administrators swaying amid the wind of pressure groups.

The independence of local boards and districts makes coordinating reform efforts a daunting task. Hildebrand (1992) compares the American-style school board to a domestic station wagon—it won't sell abroad! In other nations local control is seen as inherently unfair (we'd call it un-American) because it increases the discrepancies in opportunity between students in different districts.

The Public Agenda Foundation (1993) reports significant gaps in communication among school personnel. In a study of four districts, they found that teachers learned more from the press about district policy changes than from the educational leadership, and that conversations on these issues were dominated by rumor and speculation.

The task is challenging, but the Standards invite teachers, school administrations, and school boards to work together toward change. We have the background and knowledge to be major participants in this effort. Here are some practical steps on how to begin:

SYSTEM STANDARD B

Policies that influence science education should be coordinated within and across agencies, institutions, and organizations.

Reprinted with permission from the *National Science Education Standards.* © 1996 National Academy of Sciences. Courtesy of the National Academy Press, Washington, D.C.

Make Reform a Priority. School reform takes time. Hildebrand suggests that very few efforts at school-based management succeed because teachers and parents don't want to devote the time it takes to manage the process well, and administrators don't want to spend the time exploring the implications of change.

Start with a small investment. Insist that 10 minutes of each faculty meeting be spent discussing a "big issue" (like the National Science Education Standards). At first the clock-watchers may get

156   Navigating the System Standards

impatient, but the possibility of becoming involved in decisions that affect one personally can be appealing.

Move on with other teachers as a team to school board meetings. Stand at the microphone. Ask the board members to shorten the business agenda and devote some of *their* time each month to discuss a substantive issue with teachers.

Change the Players. When it comes time to select a new principal or superintendent, get involved in the process. A recent RAND study suggests that the best superintendents are visionary, hands-off individuals who build coalitions among community stakeholders. But most school boards tend to select a micromanager or business expert. Very few administrative candidates have a background in science education, but many are good educators. Take a stand.

Change the Priorities. Beyond the selection process, we as teachers can become involved in changing the job description of administrators. What part of an administrator's job could we (or nonteaching staff) assume in exchange for some help in changing the curriculum, networking, or improving facilities? Try one of these references for inspiration: Sloan (1993) describes a unique project in science enrichment coordinated by an elementary principal; Donivan (1993) describes a partnership with the building principal that led to success; and Halpin (1992) describes the support a principal can give when he or she speaks the same language as the faculty.

Share your vision. Don't immediately assume that your principal, superintendent, or board members understand your viewpoint or even plan on taking the time to learn about it. Buy them a cup of coffee, and take time to talk with them. Or just slip the letter to an administrator (Texley, 1991) under her or his door.

Parallax compares different views, but ends at the same point. Our views usually don't have to contrast with those of the administration; they can complement one another. Be an advocate for your school and your system. We can move together toward the Standards—or the stars.

Resources for the Road

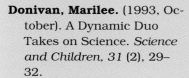

Donivan, Marilee. (1993, October). A Dynamic Duo Takes on Science. *Science and Children, 31* (2), 29–32.

Halpin, Robert. (1992, April). At Tulip Grove, A Principal for Science. *Science and Children, 29* (7), 34–35.

Hildebrand, John. (1992, Fall). Local vs. National. *America's Agenda: Schools for the 21st Century,* 27–30.

Lewis, Anne C. (1991, Spring). Sharing Power. *America's Agenda: Schools for the 21st Century, 1* (1), 40–43.

Public Agenda Foundation. (1993). *Divided Within; Besieged Without: The Politics of Four American School Districts.* New York: Author.

Sloan, Gayle. (1993, May). This Principal's Interest Is Science. *Science and Children, 30* (8), 19–20.

Texley, Juliana. (1991, February). Editor's Corner: Science Education. *The Science Teacher, 58* (2), 6.

U.S. Department of Education, Office of Educational Research and Improvement. Who Runs the Schools? The Principal's View (1993, May). The Teacher's View (1993, June). (Education Consumer Guides). Washington, DC: Author.

Welch, Wayne. (1978). Science Education in Urbanville: A Case Study. In R. Stake and J. Easley (Eds.), *Case Studies in Science Education.* Urbana, IL: University of Illinois.

About System Standard C

Continuity

Policies must be sustained over time so that significant change can be measured against the criteria we have established.

The Pendulum Experiment

Most teachers' careers last for three or more decades, and in that time it's not uncommon to see the pendulum of educational trends swing from conservative to liberal and back again. Society sees itself in its schools; and just as social trends change, so do expectations of teachers and classrooms. The results of elementary and secondary education programs aren't seen until students enter adult careers, and by that time the public often has changed its definition of "good schools." Given the swinging pendulum, how can we energize ourselves about this latest round of reform—the National Science Education Standards?

Marc Tucker (1992) suggests that schools are really "test preparation organizations"—at least to some people. But what is the test? Is it on paper? On the job? In college? Or the challenge of the polling booth? Tucker argues that setting standards will never change schools until we all agree on the test.

He suggests that the message of setting standards is that we will banish the bell curve. This means that as a nation, we will abandon the belief that only a portion of the population will understand science. Yet even now a significant social movement of conservative voters is massing against heterogeneous grouping in favor of again tracking students through schools.

In a very important sense, the National Science Education Standards are very different than the Sputnik-era reforms or the back-to-basics movement. The difference lies in how they were developed—by consensus-building by scientists, business, industry, legislators, teachers, administrators, and parents. Because *all* stakeholders were involved in developing the vision for *all* students of science, there is hope these

SYSTEM STANDARD C

Policies need to be sustained over sufficient time to provide the continuity necessary to bring about the changes required by the *Standards*.

Reprinted with permission from the *National Science Education Standards*. © 1996 National Academy of Sciences. Courtesy of the National Academy Press, Washington, D.C.

Standards will carry us well into the 21st century.

To focus education systems on more permanent change, we must begin to think in terms of long-range goals. We need to look at the long-term validity of the reforms we make in our schools and systems and how short-term indices relate to those gains.
- If our goal is employability, how is success in school science related? Do we evaluate communication skills? Informational reading?

158   Navigating the System Standards

Cooperative discovery? Ability to follow directions?
- If our goal is decisionmaking ability by tomorrow's citizens, do we practice and evaluate real-world choices in schools? Are these activities included in the grading process?
- If our goal is problem solving and logical reasoning, do we provide real-world situations in which students can practice solving problems?

A systemic approach to change must also include the major high-stakes tests that are unavoidable in today's society. It doesn't help to deny they exist or to downgrade their importance. Rather, we need to consider how we can make them better. For example, many states have moved to more relevant statewide achievement measures. Our schools can help by serving as pilot sites and giving input in the test-development process.

Finally, to achieve continuity in our programs we must show our students that we *expect* that these Standards are here to stay. We must focus our students' sights as well as our own on the long haul. Memorizing for tomorrow's test won't do it; practicing for the future will.

Can society escape the pendulum ride that education has been on for the past 90 years? Perhaps, but it will take a partnership comprising *every* part of the system to make it happen. The vision must begin in our classrooms and move through our communities to our state capitals and Washington, D.C. Together, can we see that far?

Resources for the Road

Association for Supervision and Curriculum Development (ASCD). (1995, June). Responding to Public Opinion. *Education Update, 37* (5), 1, 2–6.

Ledell, Marjorie, and Arnsparger, Arleen. (1993). *How to Deal with Community Criticism of School Change.* Alexandria, VA: Association for Supervision and Curriculum Development (ASCD).

Tucker, Marc. (1992, Fall). Measuring Up. *America's Agenda: Schools for the 21st Century,* 21–22.

About System Standard D

Resources

Policies must be supported with resources. Systemic reform often requires identifying hidden resources that can be tapped.

The Hidden Resource in Every School

The Pentagon never has bake sales—and schools almost never have enough financial resources to develop new educational "weapons." It's incongruous that a system on which the nation's future depends is supported by do-or-die local tax referendums and daily nickel counting. But the limited financial base of our schools has been a fact of life for generations and isn't likely to change soon.

Where can schools that want to move toward the Standards turn for support? When lines of communication are good, the board of education will do what it can (see System Standard A). Where opportunities exist, business and industry will lend a hand (see System Standard B). But for real, day-to-day support, most of us will have to dig for gold much closer to home.

In every school the hidden resource is parents—those qualified, sincere supporters who lurk by the school doors at dismissal time, sell candy, and buy raffle tickets. The potential of parents as partners in schools is almost always underestimated. We don't readily accept external observers into our classrooms, and the sporadic antagonistic relationship between a teacher and a dissatisfied parent can haunt efforts to encourage positive involvement.

But parent involvement is a gold mine in every community. When parents are involved in projects with schools and communities, partnerships are forged that not only maximize resources but also increase support for schools. Consider these examples:

Mentoring. Sylvania, Ohio, schools pair students with parents and representatives of the University of Toledo to develop independent projects (DeBruin et al, 1993).

Contracting Achievement. In Agana, Guam, contracts among parents, students, and teachers encourage science activities. Parents support outside activities and monitor learning logs (Mafnas et al, 1993).

Community Facilities. Libraries and museums can be invaluable partners in thematic programs. Salt Lake City schools have expanded their walls by forming a long-term partnership with museums (Greenhalgh, 1995).

Educating Parents. In Union, New Jersey, teachers contribute to the success of students by educating parents through "Parent University" (Arnold and Birne, 1995). Many other books and publications that can help parents and students discover science are available through NSTA and local bookstores.

SYSTEM STANDARD D

Policies must be supported with resources.

Reprinted with permission from the *National Science Education Standards*. © 1996 National Academy of Sciences. Courtesy of the National Academy Press, Washington, D.C.

Schools Serving Families. In most communities the school is the most identifiable government building. Schools that house family services (medical clinics, vocational libraries, social work services, and the like) give parents reasons to appreciate—and later support—their schools (Elkind, 1995).

Equipment and Facilities. What is dated at IBM is postmodern to most schools. Many businesses rotate equipment each year and appreciate the opportunity to write it off as a donation to schools. Try to find a parent who can serve as a contact (Winston, 1995).

Parents can help manage equipment, too. Most of the problems with lab science involve storage, the dispensing of chemicals, alertness for safety hazards—all tasks that parents can and like to do.

Using Technology. Your municipality's contract with the local cable company probably contains a provision to give your school—and the parents—access to the airwaves. Consider making a video to show what you are doing or to show what sort of help you need. Find a high school student who knows video editing.

Also involve parents in supervising home TV watching (if you can't beat it, join it). Share videotapes or audiotapes for use at home, and develop a parent guide to direct discussions about the tapes (McCormack, 1990).

This list is only a beginning. Youngsters learn 18 hours a day—not six. If we enlist the help of their other teachers—their parents—we'll strike gold in our schools and communities.

Resources for the Road

Arnold, Lois B., and Birne, Adriana M. (1995, November). Parent University. *The Science Teacher, 62* (8), 29–31.

Daisey, Peggy, and Shroyer, M. Gail. (1995, November/December). Parents Speak Up. *Science and Children, 33* (3), 24–26.

DeBruin, Jerry, Boellner, Carolyn, Flaskamp, Ruth, and Sigler, Karen. (1993, March). Science Investigations Mentorship Program. *Science and Children, 30* (6), 20–22.

Elkind, David. (1995, September). The School and Family in the Postmodern World. *Kappan, 77* (1), 8–14.

Greenhalgh, Lanai. (1995, September). Youth Teaching Youth. *Science and Children, 33* (1), 35–36, 75.

Mafnas, Irene, Flis, Julie Calvo, and Dionio, Suzanne. (1993, September). A Contract for Science. *Science Scope, 17* (1), 45–48.

McCormack, Alan J. (1990, October). The Family Channel. *Science and Children, 28* (2), 24–26.

North Carolina Museum of Life and Science. (1993). *Sharing Science with Children: A Guide for Parents.* Durham, NC: Author.

Partnership Possibilities [series of articles in Special Issue: Science in Nontraditional Settings]. (1995, March). *Science Scope, 18* (6), 5–28.

Paulu, Nancy, and Martin, Margery. (1992). *Helping Your Child Learn Science.* Washington, DC: U.S. Department of Education, OERI.

Pearlman, Susan, and Perica-Spector, Kathleen. (1993, November/December). A Science Open House. *Science and Children, 31* (3), 12–15.

Pearlman, Susan, and Perica-Spector, Kathleen. (1992, April). Helping Hands from Home. *Science and Children, 29* (7), 12–14.

Sussman, Art (Ed.). (1993). *Science Education Partnerships: Manual for Scientists and K–12 Teachers.* San Francisco: University of California.

U.S. Department of Education. (1995). *An Invitation to Your Community: Building Community Partnerships for Learning.* Washington, DC: Author.

Winston, H. Michael. (1995, April). Community Collaboration. *The Science Teacher, 62* (4), 20–22.

About System Standard E

Equity

Policies must support equity for all students.

Evening the Odds

One of the by-products of local control of schools is that support for schools is inherently unequal. Schools differ not only in financial resources but also in the vastly different experiential foundations of their students. These differences are related to

- economic resources, including access to food, clothing, technology, travel, and exploration (Forty-three percent of African American children, 35 percent of Hispanic children, and 14 percent of white children live in poverty in America and, therefore, are likely to attend schools that are poorly funded.)
- transience (moving from place to place or home to home)
- cultural diversity, with the accompanying differences in language, values, and perspective
- family stability and stability of child care (Fifty-four percent of African American children, 28 percent of Hispanic children, and 17 percent of white children live in single-parent families.)
- language abilities (in English and in other languages)

From the day they enter kindergarten, children reveal the disparity in their development and in their life experiences. To learn science, they must start from the world they know, using the abilities they have, and be allowed to actively explore the natural world. Young children then will learn to communicate what they know. We must be alert to developmental and experiential delays in order to help children overcome them.

> **SYSTEM STANDARD E**
>
> Science education policies must be equitable.
>
> Reprinted with permission from the *National Science Education Standards*. © 1996 National Academy of Sciences. Courtesy of the National Academy Press, Washington, D.C.

Inadequate nutrition and untreated health problems, such as ear infections, for example, can cause delayed development. And studies clearly show that differences in the ability to express ideas at age five are closely correlated to differences in logical abilities and learning in the years that follow.

The issue of equity has major legal implications. Kansas District Court Judge Terry L. Bullock sent ripples through other state courts when he issued a consolidation decision that mandated "different expenditures in different places" to ensure equal educational opportunity for children with different backgrounds. The implication:

More funds should go to schools where students have a lower "opportunity to learn" outside the classroom walls (Kozol, 1992).

To create equal opportunity to learn, school systems must rely on *compensatory education*. In the past, special programs have often excluded or separated low achievers, but research and best practice have shown that to be counterproductive. Separating and labeling students often exacerbates low achievement, limits cooperative learning, and eliminates student role models for behavior. Instead, today's schools are using co-teaching and technology to provide the extra help students need to leap forward.

Co-teaching is a system in which a special-needs teacher regularly works side-by-side with a general education teacher. The co-teacher is not an aide to the teacher but offers special skills in modification and diagnosis to the classroom. When the co-teacher is absent, a substitute is provided—an administrative acknowledgment of the importance of the partnership. Co-teaching has the advantage of enabling special-needs personnel to serve all students, not only those with special labels.

Technology now offers artificial intelligence systems to help students raise their achievement levels. Computer software helps diagnose and remediate individual students, and it keeps a record of their achievements and deficiencies. Because computer programs have endless patience and are nonjudgmental, they are appropriate supplements. Technology can also extend the school day by providing compensatory education that supplements rather than supplants the normal school program and is available for parents to explore with their children after hours.

One of the major forces for compensatory education in the United States is the Title I program. Once limited to reading and mathematics, Title I now allows schools to implement their own site-based programs to support learning in any area. Many schools are using science as an avenue to increase communication and problem-solving skills in all students.

(*Note:* For information on equity issues, see Teaching Standard E, page 19; Assessment Standard B, page 49; and Program Standards E, page 143.)

Resources for the Road

Allen-Sommerville, Lenola. (1996, February). Capiltalizing on Diversity. *The Science Teacher*, 63 (2), 20–23.

Baptiste, H. Prentice, and Key, Shirley Gholston. (1996, February). Cultural Inclusion: Where Does Your Program Stand? *The Science Teacher*, 63 (2), 32–35.

Bernhardt, Elizabeth, Hirsch, Gretchen, Teemant, Annela, and Rodriguez-Munoz, Marisol. (1996, February). Language Diversity and Science. *The Science Teacher*, 63 (2), 25–27.

Carey, Shelley Johnson (Ed.). (1993). *Science for All Cultures*. Arlington, VA: National Science Teachers Association (NSTA).

The Commission on Chapter I. (1992). *Making Schools Work for Children in Poverty*. Washington, DC: American Association for Higher Education (AAHE).

Johnston, Donald H. (1992, Winter). Readiness: Vague Idea, Pressing Need. *America's Agenda: Schools for the 21st Century*, 29–31.

Kahle, Jane Butler. (1985). *Women in Science*. Philadelphia: Falmer.

Kozol, Jonathan. (1992, Winter). Into the Mainstream. *America's Agenda: Schools for the 21st Century*, 23–24.

Stevens, Floraline I., and Grymes, John. (1993). Opportunity to Learn: Issues of Equity for Poor and Minority Students. Washington, DC: U.S. Dept. of Education.

About System Standard F

Unanticipated Effects

Policies must be examined for possible unintended effects on classroom practice.

Ripples in a Pond

Can we track the path of reform? Educators usually begin with a plan that looks linear, but the results often resemble a child's concept map—a splatter of effects and side effects, only a few of which were intended or anticipated. Educational reformers craft short- and long-term outcomes for students and then prescribe changes to achieve them. In creating policy, we must be prepared for both intended and unanticipated effects no matter how careful our plans. Here are some examples:

- In the late 1980s, the National Science Foundation convened a panel to study the long-term effects of the Presidential Awards program. To its dismay, the panel found that after five years a significant number of the 1983 and 1984 classes of the nation's best science teachers had left the classroom for administration and consulting. The Presidential Awards were created to recognize great teaching in the classroom. Was the unanticipated result a negative? Or did the principles of good teaching get a wider audience?
- Twenty-two states now have statewide achievement testing in science. These policies were supposed to increase the amount of time spent on school science; but the unanticipated effects have been an emphasis on test-taking skills and a de-emphasis on higher-order thinking.
- Increased consciousness of safe storage and disposal of chemicals was intended to make laboratory science safer for elementary students. But it has also reduced the number of elementary teachers who are willing to do chemistry experiments at all.

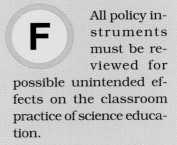

SYSTEM STANDARD F

All policy instruments must be reviewed for possible unintended effects on the classroom practice of science education.

Reprinted with permission from the National Science Education Standards. © 1996 National Academy of Sciences. Courtesy of the National Academy Press, Washington, D.C.

How can policymakers minimize negative unanticipated effects? The only clear answer is consistent evaluation. They need to work closely with teachers because we are the ones who see what happens day by day. Policies can't be put into place and then abandoned to schools and teachers to implement. The key to consistency is process: beginning, middle, and end.

In 1992 NSTA coordinated a symposium on change in schools. Through the support of the Monsanto Fund, NSTA launched an effort to help local school district teams try to implement change. The process they used (documented in NSTA's *A Strategy for Change*) is a model for designing systemic reform that incorporates continual monitoring and internal assessment for unanticipated effects.

The NSTA project began by soliciting volunteer teams of teachers and administrators from 175 school districts. Thirty-eight teams participated in the symposium. Many of the districts had already begun working on reform, but all were ready for a coordinated self-assessment. The work of the teams emphasized the practical side of systemic reform, and participants were constantly encouraged to look for barriers and unanticipated effects. For example, the teams identified early on what was *needed* to implement change. Then they identified what *hindered* change:

- time wasted talking and not doing
- slow-to-change textbook publishers
- little training for teachers in materials management
- not enough preparation time in the classroom
- lack of ways to circumvent barriers in school systems

These insights indicate the value of teacher-based planning. The real barriers to change go far beyond standards and assessments—to daily living.

NSTA interviewed the teams one year after the symposium. Most school districts had experienced many predicted effects and could cite positive changes. But many also reported events they had not anticipated:

- San Diego Public Schools planned with vigor—then experienced frustration when a $28 million budget cut hit their district.
- Greater Albany, Oregon, saw its entire secondary curriculum structure change, divided into humanities and sciences.
- Anchorage, Alaska, compared its efforts to the ripples formed by a small pebble in a calm lake. The entire community had changed from the team's work.

We should keep the Anchorage analogy of ever-widening ripples in a pond in mind. As innovators and educators, we must remember that a single change in school science will affect many other areas in ways none of us can predict. We must be sensitive to changes, predicted and unanticipated, and keep our antennae out to sense them. We must be prepared to react and adjust when our intended course changes. Only then can we move confidently toward the Standards.

Resources for the Road

ERIC, U.S. Department of Education. Systemic Educational Reform (Theme Issue). (1994, Fall). *ERIC Review, 3* (2), 1–28.

Institute for Education in Transformation, Claremont Graduate School. *Voices from the Inside: A Report on Schooling from Inside the Classroom.* (1992). Claremont, CA: Author.

McGonagill, Grady. (1993). Overcoming Barriers to Educational Restructuring: A Call for System Literacy (ERIC Document Reproduction Service No. ED357512).

National Science Teachers Association. (1993). *A Strategy for Change in Elementary School Science: Proceedings of Conference.* Arlington, VA: Author.

About System Standard G

Individual Responsibility

Responsible individuals will take the opportunity presented by the Standards to move their systems toward reform.

The Wings of a Single Butterfly

With so many components in the education system, it's easy to feel powerless. But we can take a lesson from modern mathematical chaos theory: When a single butterfly flaps its wings, air currents around the world are affected. In the same way, every contribution, no matter how small, influences the whole. As science teachers, we must never consider the enormity of the task an excuse not to begin reform. There are simple, sound ways to foster change in *your* system.

In many districts, administrations and organizations of teachers are joining together for activism. Traditionally antagonistic union negotiations are becoming more rare as stakeholders realize the benefits of win-win cooperation. Conley (1994) suggests that before any school district begins reform, it should consider the following questions:

- Are efforts being made to include the professional association (union) as a partner?
- Are bargaining options being explored at the district level?
- Are good-faith efforts being made to redefine the role of the union?
- Are strategies being developed to create new collaborations between administration and teachers?

Once teachers and schools have joined together for local change, efforts to change the system must reach beyond individual schools. Resisting the impulse to shut our classroom door, we must move out into the political arena if changes we seek are to become permanent.

All over the nation teachers are running for school boards and legislatures. They are writing and speaking about their profession. And they are using the media to tell the story of school science from their own, unique perspective. This is the only way real change will occur.

> **SYSTEM STANDARD G**
>
>
>
> Responsible individuals must take the opportunity afforded by the standards-based reform movement to achieve the new vision of science education portrayed in the *Standards*.
>
> Reprinted with permission from the *National Science Education Standards*. © 1996 National Academy of Sciences. Courtesy of the National Academy Press, Washington, D.C.

When It's Our Turn

Eventually it will come to a system near each of us—an issue, a controversy, or the opportunity to make a significant change in the way science is taught in our school. The challenge may come from a pressure group, from parents, or, ideally, from the message of the National Science Education Standards. Will we be part of the process?

As individual teachers, working within the system, the key to our effectiveness will be networking. We are never alone in our efforts. We should use the resources our profession provides:

Step 1: Rely on Research. Share this book and others with those who will make decisions. NSTA also offers an awareness kit for teachers on the National Science Education Standards; another kit is available for administrators.

Step 2: Bring Friends Along. Convince your colleagues of the importance of activism. Share your information and your insights.

Step 3: Network on the Net. Access teacher forums (NSTA home page, America Online, Compuserve, Prodigy, or the Internet) for quick answers to difficult questions in systemic reform.

Step 4: Maintain Your Memberships. Many of us feel the main advantage of professional memberships are the publications. But a major portion of the activities of professional organizations is speaking out on issues important to members. Keep up with what is happening.

Step 5: Call in the Militia. If an issue comes before your school board that is so detrimental or so difficult to combat that it threatens good science education, send out a call for help. Your state science supervisor, your NSTA District or Division Director, or your professional association can find you expert help in your area.

It doesn't take a hurricane to move the atmosphere—even the flap of a single butterfly's wings has an effect across the globe. So the involvement of a single teacher can change the system. And given that, just imagine the impact of our entire profession speaking with one voice!

Resources for the Road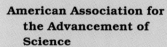

American Association for the Advancement of Science
1333 H St. NW
Washington, DC 20005

Annenbery/CPB Math and Science Project
901 E St. NW
Washington, DC 20004

Center for Educational Reform
1001 Connecticut Ave. NW
Suite 920
Washington, DC 20036

Center for Systemic School Reform
San Francisco State University
221 Burk Hall
1600 Holloway Ave.
San Francisco, CA 94132

Coalition of Essential Schools
Brown University
Box 1969
Providence, RI 02912

Education Commission of the States
707 17th St.
Suite 2700
Denver, CO 80202

National Center for Restructuring Education, Schools, and Teaching
Teachers College
Columbia University
Box 110
New York, NY 10027

National Governors' Association
444 North Capitol St.
Suite 267
Washington, DC 20001

National Science Foundation Office of Systemic Reform
4201 Wilson Blvd.
Arlington, VA 22230

National Science Teachers Association
1840 Wilson Blvd.
Arlington, VA 22201

New Standards Project
700 11th St. NW
Suite 750
Washington, DC 20001

Changing Emphases
District System

LESS EMPHASIS ON	MORE EMPHASIS ON
Technical, short-term, inservice workshops	Ongoing professional development to support teachers
Policies unrelated to Standards-based reform	Policies designed to support changes called for in the Standards
Purchase of textbooks based on traditional topics	Purchase or adoption of curriculum aligned with the Standards and on a conceptual approach to science teaching, including support for hands-on science materials
Standardized tests and assessments unrelated to Standards-based program and practices	Assessments aligned with the Standards
Administration determining what will be involved in improving science education	Teacher leadership in improvement of science education
Authority at upper levels of educational system	Authority for decisions at level of implementation
School board ignorance of science education program	School board support of improvements aligned with the Standards
Local union contracts that ignore changes in curriculum, instruction, and assessment	Local union contracts that support improvements indicated by the Standards

Reprinted with permission from the *National Science Education Standards*. © 1996 National Academy of Sciences. Courtesy of the National Academy Press, Washington, D.C.

Appendices

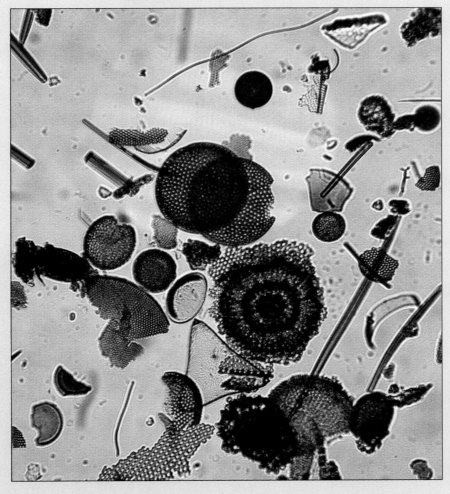

Appendix A
National Science Education Standards: Some Relevant History

The publication of the *National Science Education Standards* in December 1995 propelled science education into a new and challenging era. Never before has the practice of science education in the United States been guided by a single set of principles reached by national consensus. The *National Science Education Standards* take their place as a significant contribution to the broad reform movement currently under way in American education.

Precursors to Reform

In April 1983, *A Nation at Risk*, thought to be the most important reform publication of this century, warned that if our education system continued to produce citizens illiterate in science, mathematics, and technology, the nation would lose its influential position in the world, becoming a second-rate power in the 21st century. *A Nation at Risk* sparked a wealth of studies and other efforts that eventually coalesced into a significant broad-based reform movement in education.

Among the first studies were efforts to compare the literacy levels of U.S. students with those of students in other countries. Although many studies were poorly designed, making their findings invalid, the better studies consistently found a need to improve the teaching of science and mathematics in this country.

Launching Reform

Among the various efforts to reform science education was Project 2061, initiated in 1986 by the American Association for the Advancement of Science. This project takes a long-term view of science education reform. The project's goal is to develop a high level of science literacy among all American citizens. Its first publication, *Science for All Americans* (1989), outlined the understandings and habits of mind necessary for a scientifically literate citizen. In 1993, Project 2061 released *Benchmarks for Science Literacy*, which established minimum goals for what students should know and be able to do at various grade levels in a number of content areas.

In 1989, NSTA launched its Scope, Sequence, and Coordination of Secondary School Science project (SS&C). The NSTA curriculum reform project sought to revamp the layercake approach to the study of science (a different science every year). Instead, a curriculum based on the principles of SS&C would give students carefully sequenced (from concrete to abstract, paralleling student development), well-coordinated instruction in *all* the sciences *every* year. The SS&C approach is currently being field-tested at the high school level. Previous efforts focused on the middle grades. Publications of this project include *The Content Core, Relevant Research*, and *A High School Framework for Science Education*.

Goals and Standards

While these efforts were going on, two other events were propelling the nation toward setting standards for school subjects. In 1989, after three years of work, the National Council of Teachers of Mathematics released its *Curriculum and Evaluation Standards for School Mathematics*. Its goal was to revolutionize the teaching of mathematics as a subject for *all* students, not just those who were college-bound.

About the same time President Bush decided to convene an education summit of the nation's governors. From this summit came an agreement that national goals for education should be set. The National Governors' Association and the President developed the National Education Goals, which were released in the State of the Union address in early 1990. According to Goal 4, "By the year 2000, U.S. students will be first in the world in science and mathematics achievement."

The National Education Goals Panel was established to monitor progress toward each of the goals. President Bush launched the America 2000 effort to get communities involved in working toward the goals. (Later, President Clinton would continue these goal-directed efforts, renaming the program Goals 2000.)

The need to set *voluntary* national standards for what all students should know and be able to do at various grade levels in the subjects addressed by the goals soon became apparent. These included science, mathematics, English, history, civics and government, geography, economics, foreign languages, and the arts. Standards-setting projects, often with funding from agencies of the federal government, were quickly launched in these subjects, including science.

Toward Consensus in Science

In spring 1991, the president of NSTA, with the unanimous support of the NSTA board of directors, wrote to Frank Press, president of the National Academy of Sciences and chair of the National Research Council (NRC), asking NRC to coordinate the development of national standards in science education. The presidents of other leading science and science education associations, the U.S. Secretary of Education, the assistant director for education and human resources at the National Science Foundation (NSF), and the co-chairs of the National Education Goals Panel also encouraged NRC to play a leading role. NRC agreed; and shortly thereafter, major funding for the project was provided by the Department of Education and NSF. Other funders included NASA, the U.S. Department of Energy, the U.S. Department of Agriculture, and the National Institutes of Health.

NRC began its work by convening the National Committee on Science Education Standards and Assessment and the Chair's Advisory Committee, consisting of representatives from the major science and science education associations. This group helped to identify and to recruit staff and volunteers for all the standards-writing working groups.

The three working groups—content standards, teaching standards, and assessment standards—set to work in summer 1992 and over the next 18 months, solicited input from large numbers of teachers, scientists, science educators, and many others interested in science education.

In spring 1993, a pre-draft of the science education standards was released to selected focus groups for critique and review. Comments were collated and analyzed; and in December 1994, more than 40,000 copies of the draft *National Science Education Standards* were distributed nationwide to some 18,000 individuals and 250 groups. Again, comments were collated and analyzed, and the *National Science Education Standards* was published in December 1995.

The Science Standards provide a vision, *not* a curriculum, for science education. They are descriptive, *not* prescriptive. One of the strongest principles underlying the Standards is that science is for *all* students in all grades.

Release of the *National Science Education Standards* is a pivotal event for teachers of science and those they work with—from students and parents to administrators and legislators. The Standards present clearly what *can* be done but leave the nuts and bolts of implementation to individual choice and responsibility. The *Standards* clearly state that all the standards should be taken together as a package—implementation should not be done cafeteria-style. And finally, the responsibility for putting the vision of the Standards into action belongs to *everyone* in science education: teachers, curriculum developers, superintendents, administrators, supervisors, policymakers, assessment specialists, scientists, teacher educators, parents, businesses, and local communities.

Resources for the Road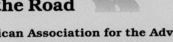

American Association for the Advancement of Science (AAAS), Project 2061. (1989). *Science for All Americans.* New York: Oxford University Press.

American Association for the Advancement of Science (AAAS), Project 2061. (1993). *Benchmarks for Science Literacy.* New York: Oxford University Press.

Aldridge, Bill G. (Ed.) (1996). *Scope, Sequence, and Coordination: A High School Framework for Science Education.* Arlington, VA: National Science Teachers Association (NSTA).

National Commission on Excellence in Education. (1983). *A Nation at Risk: The Imperative for Educational Reform.* Washington, DC: Author.

National Council of Teachers of Mathematics (NCTM). (1989). *Curriculum and Evaluation Standards for School Mathematics.* Reston, VA: Author.

National Council of Teachers of Mathematics (NCTM). (1989). *Professional Standards for Teaching Mathematics.* Reston, VA: Author.

National Science Teachers Association (NSTA) (1993). *Scope, Sequence and Coordination of Secondary School Science. Vol. I. The Content Core: A Guide for Curriculum Designers.* (Rev. ed.). Arlington, VA: Author.

National Science Teachers Association (NSTA). (1992). *Scope, Sequence, and Coordination of Secondary School Science. Vol. II. Relevant Research,* Arlington, VA: Author.

Appendix B
National Science Education Standards

Principles

- Science is for all students.
- Learning science is an active process.
- School science reflects the intellectual and cultural traditions that characterize the practice of contemporary science.
- Improving science education is part of systemic education reform.

Science Teaching Standards

TEACHING STANDARD A: Teachers of science plan an inquiry-based science program for their students. In doing this, teachers
- develop a framework of year-long and short-term goals for students
- select science content and adapt and design curricula to meet the interests, knowledge, understanding, abilities, and experiences of students
- select teaching and assessment strategies that support the development of student understanding and nurture a community of science learners
- work together as colleagues within and across disciplines and grade levels

TEACHING STANDARD B: Teachers of science guide and facilitate learning. In doing this, teachers
- focus and support inquiries while interacting with students
- orchestrate discourse among students about scientific ideas
- challenge students to accept and share responsibility for their own learning
- recognize and respond to student diversity and encourage all students to participate fully in science learning
- encourage and model the skills of scientific inquiry, as well as the curiosity, openness to new ideas and data, and skepticism that characterize science

TEACHING STANDARD C: Teachers of science engage in ongoing assessment of their teaching and of student learning. In doing this, teachers
- use multiple methods and systematically gather data about student understanding and ability
- analyze assessment data to guide teaching
- guide students in self-assessment
- use student data, observations of teaching, and interactions with colleagues to reflect on and improve teaching practice
- use student data, observations of teaching, and interactions with colleagues to report student achievement and opportunities to learn to students, teachers, parents, policymakers, and the general public

Reprinted with permission from the *National Science Education Standards*. © 1996 National Academy of Sciences. Courtesy of the National Academy Press, Washington, D.C.

TEACHING STANDARD D: Teachers of science design and manage learning environments that provide students with the time, space, and resources needed for learning science. In doing this, teachers
- structure the time available so that students are able to engage in extended investigations
- create a setting for student work that is flexible and supportive of science inquiry
- ensure a safe working environment
- make the available science tools, materials, media, and technological resources accessible to students
- identify and use resources outside the school
- engage students in designing the learning environment

TEACHING STANDARD E: Teachers of science develop communities of science learners that reflect the intellectual rigor of scientific inquiry and the attitudes and social values conducive to science learning. In doing this, teachers
- display and demand respect for the diverse ideas, skills, and experiences of all students
- enable students to have a significant voice in decisions about the content and context of their work and require students to take responsibility for the learning of all members of the community
- nurture collaboration among students
- structure and facilitate ongoing formal and informal discussion based on a shared understanding of rules of scientific discourse
- model and emphasize the skills, attitudes, and values of scientific inquiry

TEACHING STANDARD F: Teachers of science actively participate in the ongoing planning and development of the school science program. In doing this, teachers
- plan and develop the school science program
- participate in decisions concerning the allocation of time and other resources to the science program
- participate fully in planning and implementing professional growth and development strategies for themselves and their colleagues

Standards for Professional Development for Teachers of Science

PROFESSIONAL DEVELOPMENT STANDARD A: Professional development for teachers of science requires learning essential science content through the perspectives and methods of inquiry. Science learning experiences for teachers must
- involve teachers in actively investigating phenomena that can be studied scientifically, interpreting results, and making sense of findings consistent with currently accepted scientific understanding
- address issues, events, problems, or topics significant in science and of interest to participants
- introduce teachers to scientific literature, media, and technological resources that expand their science knowledge and their ability to access further knowledge
- build on the teacher's current science understanding, ability, and attitudes
- incorporate ongoing reflection on the process and outcomes of understanding science through inquiry
- encourage and support teachers in efforts to collaborate

PROFESSIONAL DEVELOPMENT STANDARD B: Professional development for teachers of science requires integrating knowledge of science, learning, pedagogy, and students; it also requires applying that knowledge to science teaching. Learning experiences for teachers of science must
- connect and integrate all pertinent aspects of science and science education
- occur in a variety of places where effective science teaching can be illustrated and modeled, permitting teachers to struggle

with real situations and expand their knowledge and skills in appropriate contexts
- address teachers' needs as learners and build on their current knowledge of science content, teaching, and learning
- use inquiry, reflection, interpretation of research, modeling, and guided practice to build understanding and skill in science teaching

PROFESSIONAL DEVELOPMENT STANDARD C: Professional development for teachers of science requires building understanding and ability for lifelong learning. Professional development activities must
- provide regular, frequent opportunities for individual and collegial examination and reflection on classroom and institutional practice
- provide opportunities for teachers to receive feedback about their teaching and to understand, analyze, and apply that feedback to improve their practice
- provide opportunities for teachers to learn and use various tools and techniques for self-reflection and collegial reflection, such as peer coaching, portfolios, and journals
- support the sharing of teacher expertise by preparing and using mentors, teacher advisors, coaches, lead teachers, and resource teachers to provide professional development opportunities
- provide opportunities to know and have access to existing research and experiential knowledge
- provide opportunities to learn and use the skills of research to generate new knowledge about science and the teaching and learning of science

PROFESSIONAL DEVELOPMENT STANDARD D: Professional development programs for teachers of science must be coherent and integrated. Quality preservice and inservice programs are characterized by
- clear, shared goals based on a vision of science learning, teaching, and teacher development congruent with the *National Science Education Standards*
- integration and coordination of the program components so that understanding and ability can be built over time, reinforced continuously, and practiced in a variety of situations
- options that recognize the developmental nature of teacher professional growth and individual and group interests, as well as the needs of teachers who have varying degrees of experience, professional expertise, and proficiency
- collaboration among the people involved in programs, including teachers, teacher educators, teacher unions, scientists, administrators, policymakers, members of professional and scientific organizations, parents, and businesspeople, with clear respect for the perspectives and expertise of each
- recognition of the history, culture, and organization of the school environment
- continuous program assessment that captures the perspectives of all those involved, uses a variety of strategies, focuses on the process and effects of the program, and feeds directly into program improvement and evaluation

Standards for Assessment in Science Education

ASSESSMENT STANDARD A: Assessments must be consistent with the decisions they are designed to inform.
- Assessments are deliberately designed.
- Assessments have explicitly stated purposes.
- The relationship between the decisions and the data is clear.
- Assessment procedures are internally consistent.

ASSESSMENT STANDARD B: Achievement and opportunity to learn science must be assessed.
- Achievement data collected focus on the science content that is most important for students to learn.

- Opportunity-to-learn data collected focus on the most powerful indicators.
- Equal attention must be given to the assessment of opportunity to learn and to the assessment of student achievement.

ASSESSMENT STANDARD C: The technical quality of the data collected is well matched to the decisions and actions taken on the basis of their interpretation.
- The feature that is claimed to be measured is actually measured.
- Assessment tasks are authentic.
- An individual student's performance is similar on two or more tasks that claim to measure the same aspect of student achievement.
- Students have adequate opportunity to demonstrate their achievements.
- Assessment tasks and methods of presenting them provide data that are sufficiently stable to lead to the same decisions if used at different times.

ASSESSMENT STANDARD D: Assessment practices must be fair.
- Assessment tasks must be reviewed for the use of stereotypes, for assumptions that reflect the perspectives or experiences of a particular group, for language that might be offensive to a particular group, and for other features that might distract students from the intended task.
- Large-scale assessments must use statistical techniques to identify potential bias among subgroups.
- Assessment tasks must be appropriately modified to accommodate the needs of students with physical disabilities, learning disabilities, or limited English proficiency.
- Assessment tasks must be set in a variety of contexts, be engaging to students with different interests and experiences, and must not assume the perspective or experience of a particular gender, racial, or ethnic group.

ASSESSMENT STANDARD E: The inferences made from assessments about student achievement and opportunity to learn must be sound.
- When making inferences from assessment data about student achievement and opportunity to learn science, explicit reference needs to be made to the assumptions on which the inferences are based.

Science Content Standards

Content Standard: K–12

Unifying Concepts and Processes

STANDARD: As a result of activities in grades K–12, all students should develop understanding and abilities aligned with the following concepts and processes:
- systems, order, and organization
- evidence, models, and explanation
- constancy, change, and measurement
- evolution and equilibrium
- form and function

Content Standards: K–4

Science as Inquiry

CONTENT STANDARD A: As a result of activities in grades K–4, all students should develop
- abilities necessary to do scientific inquiry
- understanding about scientific inquiry

Physical Science

CONTENT STANDARD B: As a result of the activities in grades K–4, all students should develop an understanding of
- properties of objects and materials
- position and motion of objects
- light, heat, electricity, and magnetism

Life Science

CONTENT STANDARD C: As a result of activities in grades K–4, all students should develop understanding of
- the characteristics of organisms
- life cycles of organisms
- organisms and environments

Earth and Space Science

CONTENT STANDARD D: As a result of their activities in grades K–4, all students should develop an understanding of
- properties of Earth materials
- objects in the sky
- changes in Earth and sky

Science and Technology

CONTENT STANDARD E: As a result of activities in grades K–4, all students should develop
- abilities of technological design
- understanding about science and technology
- abilities to distinguish between natural objects and objects made by humans

Science in Personal and Social Perspectives

CONTENT STANDARD F: As a result of activities in grades K–4, all students should develop understanding of
- personal health
- characteristics and changes in populations
- types of resources
- changes in environments
- science and technology in local challenges

History and Nature of Science

CONTENT STANDARD G: As a result of activities in grades K–4, all students should develop understanding of
- science as a human endeavor

Content Standards: 5–8

Science as Inquiry

CONTENT STANDARD A: As a result of activities in grades 5–8, all students should develop
- abilities necessary to do scientific inquiry
- understandings about scientific inquiry

Physical Science

CONTENT STANDARD B: As a result of their activities in grades 5–8, all students should develop an understanding of
- properties and changes of properties in matter
- motions and forces
- transfer of energy

Life Science

CONTENT STANDARD C: As a result of their activities in grades 5–8, all students should develop understanding of
- structure and function in living systems
- reproduction and heredity
- regulation and behavior
- populations and ecosystems
- diversity and adaptations of organisms

Earth and Space Science

CONTENT STANDARD D: As a result of their activities in grades 5–8, all students should develop an understanding of
- structure of the Earth system
- Earth's history
- Earth in the solar system

Science and Technology

CONTENT STANDARD E: As a result of activities in grades 5–8, all students should develop
- abilities of technological design
- understandings about science and technology

Science in Personal and Social Perspectives

CONTENT STANDARD F: As a result of activities in grades 5–8, all students should develop understanding of
- personal health
- populations, resources, and environments
- natural hazards
- risks and benefits
- science and technology in society

History and Nature of Science

CONTENT STANDARD G: As a result of activities in grades 5–8, all students should develop understanding of
- science as a human endeavor
- nature of science
- history of science

Content Standards: 9–12

Science as Inquiry

CONTENT STANDARD A: As a result of activities in grades 9–12, all students should develop
- abilities necessary to do scientific inquiry
- understandings about scientific inquiry

Physical Science

CONTENT STANDARD B: As a result of their activities in grades 9–12, all students should develop an understanding of
- structure of atoms
- structure and properties of matter
- chemical reactions
- motions and forces
- conservation of energy and increase in disorder
- interactions of energy and matter

Life Science

CONTENT STANDARD C: As a result of their activities in grades 9–12, all students should develop understanding of
- the cell
- molecular basis of heredity
- biological evolution
- interdependence of organisms
- matter, energy, and organization in living systems
- behavior of organisms

Earth and Space Science

CONTENT STANDARD D: As a result of their activities in grades 9–12, all students should develop an understanding of
- energy in the Earth system
- geochemical cycles
- origin and evolution of the Earth system
- origin and evolution of the universe

Science and Technology

CONTENT STANDARD E: As a result of activities in grades 9–12, all students should develop
- abilities of technological design
- understandings about science and technology

Science in Personal and Social Perspectives

CONTENT STANDARD F: As a result of activities in grades 9–12, all students should develop understanding of
- personal and community health
- population growth
- natural resources
- environmental quality
- natural and human-induced hazards
- science and technology in local, national, and global challenges

History and Nature of Science

CONTENT STANDARD G: As a result of activities in grades 9–12, all students should develop understanding of
- science as a human endeavor
- nature of scientific knowledge
- historical perspectives

Science Education Program Standards

PROGRAM STANDARD A: All elements of the K–12 science program must be consistent with the other *National Science Education Standards* and with one another and developed within and across grade levels to meet a clearly stated set of goals.
- In an effective science program, a set of clear goals and expectations for students must be used to guide the design, implementation, and assessment of all elements of the science program.
- Curriculum frameworks should be used to guide the selection and development of units and courses of study.
- Teaching practices need to be consistent with the goals and curriculum frameworks.
- Assessment policies and practices should be aligned with the goals, student expectations, and curriculum frameworks.
- Support systems and formal and informal expectations of teachers must be aligned with the goals, student expectations, and curriculum frameworks.
- Responsibility needs to be clearly defined for determining, supporting, maintaining, and upgrading all elements of the science program.

PROGRAM STANDARD B: The program of study in science for all students should be developmentally appropriate, interesting, and relevant to students' lives; emphasize student understanding through inquiry; and be connected with other school subjects.
- The program of study should include all of the content standards.
- Science content must be embedded in a variety of curriculum patterns that are developmentally appropriate, interesting, and relevant to students' lives.
- The program of study must emphasize student understanding through inquiry.
- The program of study in science should connect to other school subjects.

PROGRAM STANDARD C: The science program should be coordinated with the mathematics program to enhance student use and understanding of mathematics in the study of science and to improve student understanding of mathematics.

PROGRAM STANDARD D: The K–12 science program must give students access to appropriate and sufficient resources, including quality teachers, time, materials and equipment, adequate and safe space, and the community.
- The most important resource is professional teachers.
- Time is a major resource in a science program.
- Conducting scientific inquiry requires that students have easy, equitable, and frequent opportunities to use a wide range of equipment, materials, supplies, and other resources for experimentation and direct investigation of phenomena.
- Collaborative inquiry requires adequate and safe space.
- Good science programs require access to the world beyond the classroom.

PROGRAM STANDARD E: All students in the K–12 science program must have equitable access to opportunities to achieve the *National Science Education Standards*.

PROGRAM STANDARD F: Schools must work as communities that encourage, support, and sustain teachers as they implement an effective science program.
- Schools must explicitly support reform efforts in an atmosphere of openness and trust that encourages collegiality.
- Regular time needs to be provided and teachers encouraged to discuss, reflect, and conduct research around science education reform.
- Teachers must be supported in creating and being members of networks of reform.
- An effective leadership structure that

includes teachers must be in place.

Science Education System Standards

SYSTEM STANDARD A: Policies that influence the practice of science education must be congruent with the program, teaching, professional development, assessment, and content standards while allowing for adaptation to local circumstances.

SYSTEM STANDARD B: Policies that influence science education should be coordinated within and across agencies, institutions, and organizations.

SYSTEM STANDARD C: Policies need to be sustained over sufficient time to provide the continuity necessary to bring about the changes required by the *Standards*.

SYSTEM STANDARD D: Policies must be supported with resources.

SYSTEM STANDARD E: Science education policies must be equitable.

SYSTEM STANDARD F: All policy instruments must be reviewed for possible unintended effects on the classroom practice of science education.

SYSTEM STANDARD G: Responsible individuals must take the opportunity afforded by the standards-based reform movement to achieve the new vision of science education portrayed in the *Standards*.

Appendix C
Designing High School Science Facilities

Program Standard D: The K–12 science program must give students access to appropriate and sufficient resources, including quality teachers, time, materials and equipment, adequate and safe space, and the community.

Proper facilities are the foundation for effective science investigations and instruction and the first line of defense in providing a safe science education environment. No curriculum, discipline, or instructional strategy can fully overcome limitations resulting from inadequate facilities.

High school students need concrete experiences that represent abstract concepts to understand scientific phenomena. The *National Science Education Standards* call time, space, and materials "critical components" for promoting sustained inquiry.

Inductive inquiry/discovery and deductive laboratory and field activities require similar facilities and equipment. The high school science room is typically a combination laboratory/classroom, but some schools provide separate laboratories.

In planning for either kind of facility, safety, flexibility, and other needs and requirements must be taken into account. The following sections offer several criteria for creating a science learning environment that encourages maximum student involvement and achievement and assists teachers in their work toward achieving the Standards.

Class Time

It is important to allot sufficient time for hands-on inquiry and activities and accompanying discussion and explanation of the science concepts involved. Plan for a minimum of 300 minutes per week of science instruction in grades 9–12, with at least 40 percent of that time devoted to inquiry/hands-on experiences.

Planning for Facilities Design

Planning involves discussion, investigation, and decisionmaking to determine the physical environment the science program requires. A participatory process that encourages input from diverse groups is most likely to result in a facility design that is specific to the district, curriculum, and program and that will have community support.

Before decisions on design and location of science facilities are made, it is important to determine exactly how science will be taught.

Participants. The planning committee for construction, addition, or renovation of high school facilities typically includes the principal, science teachers from each discipline, teachers from nonscience subjects (if the school has an integrated curriculum), repre-

sentative parents and students, the science supervisor, and the superintendent or assistant superintendent. In addition, the following groups may be represented on the committee or involved in various aspects of planning:
- science specialists
- science facilities specialists
- school administrators
- a university-based consultant
- school support services personnel
- architects
- furniture consultants
- community members
- business leaders

If the project requires specific votes or approvals by the local government, consult with appropriate officials and political leaders from the start. Include custodians and facilities maintenance staff in discussion; they have a stake in the final product and can contribute practical suggestions to ensure that the facilities can be kept in good working order long after the project is completed.

Bring special education staff members and parents of special education students into the process for their expertise in identifying accessibility issues. The district or municipality should have a specialist knowledgeable about the Americans With Disabilities Act (ADA) requirements. Consult with this person because he or she can offer suggestions on accessibility and advise architects of methods of compliance.

Preparation. Prior to beginning their work with designers, architects, and engineers, teachers and supervisors who will serve on the planning committee should try to acquire a background in understanding
- how the design and construction process works
- what factors affect construction costs
- how to read plans and specifications

A subcommittee could be formed to gather input from the community and raise community awareness. The subcommittee could develop a questionnaire and administer it to a group of students, educators, parents, and community representatives. The questionnaire should address areas of exemplary and safe science instruction as presented by the National Science Education Standards, the high school science curriculum, and state and local regulations.

Planning. The planning committee should prepare a statement of needs and educational specifications that will provide the foundation for design and development. Checklists are useful at this stage to help ensure that various building, design, funding eligibility, space, equipment, and safety requirements are met.

Determine enrollment projections for short-term, mid-term, and long-term needs. If a population surge is anticipated, determine whether the bulge can be accommodated on a temporary (5- to 10-year) basis; if so, plan facilities that allow for low-cost modification for other uses in future years. For new construction or major renovations, a 20-year projection of needs should be developed.

Planning and design considerations include the following:
- the nature of the sciences taught and the overall education program
- desired characteristics of the science facilities
- number of science facilities needed
- clustering of facilities
- relationships between facilities for science and those for other disciplines
- relationships of the science spaces to one another
- safety requirements
- furnishings and equipment
- proposed computer use and other technology needs
- outdoor education areas
- adaptations for students with disabilities
- government regulations
- maintenance requirements
- budget
- time line

Budget Priorities

The maxim "pay now or pay more later" applies here. A foundation for academic excellence begins with excellent facilities. They are not a luxury, but a requirement for maximum student achievement.

Space and safety are primary considerations in the planning budget. Lack of adequate space is much more costly to address later than the purchase of additional equipment, furniture, or new technology; also, overcrowding risks accidents and litigation. Any safety hazards resulting from poorly designed facilities will likely cause problems for the life of the building. In addition, good planning dictates that wiring and other accommodations for electronic equipment take into account both current and future needs.

In general, the cost to build and equip a stand-alone student laboratory is twice that of a regular classroom, while the cost of a combination classroom/laboratory lies between the two. For all science facilities, annual budgets need to support operating costs, equipment maintenance and acquisition, and supplies.

Space Considerations

Space is an important factor in promoting inquiry, collaborative learning, and safety. Considering current technology needs and teaching practices, a good science room will require
- a minimum of 60 ft.^{2} per pupil (5.6 m^2), which is equivalent to 1,440 ft.^{2} (134 m^2) to accommodate a class of 24 safely in a combination laboratory/classroom—or—a minimum of 45 ft.^{2} per pupil (4.2 m^2), which is equivalent to 1,080 ft.^{2} (101 m^2) to accommodate a class of 24 safely in a stand-alone laboratory (The 1990 NSTA Position Statement on Laboratory Science recommends a maximum class size of 24 students in high school.)
- additional space to accommodate students with disabilities
- 15 ft.^{2} (1.4 m^2) for each computer station with monitor, keyboard, CPU, printer, and CD-ROM
- 12 ft.^{2} (1.1 m^2) for a multimedia projector or data projector with a hard drive

Adequate space is needed in or adjacent to the classroom for longer-term student projects; a separate student project room is preferable. In addition, 10 ft.^{2} (0.9 m^2) per student is needed for teacher preparation space and for separate storage space (240 ft.^{2}, or 22 m^2, for a class of 24). Space should also be allocated for a refrigerator, dishwasher, distillation unit, and autoclave.

General Room Design

The design of an effective science room accommodates work in all science disciplines, with flexibility in furniture arrangement, abundant storage, sufficient working space for the safe conduct of activities, and holding space for ongoing projects. A rectangular room that is closer to being square works better than a long, narrow one. The room must have at least two exits and unobstructed doorways wide enough to accommodate students with physical disabilities. A ceiling height of 10 ft. (3 m) is desirable. Adequate ventilation (a minimum of four air changes per hour) is also important.

In addition to furnishings, equipment, wiring, and ventilation (discussed later), additional factors to consider are space for teacher planning, a view of the outdoors, daylight exposure (preferably southern) for plant growth, and the inclusion of a projecting plant window. Heating and cooling systems are necessary for yearround use of facilities. Utilities such as gas, vacuum, compressed air, and waste-handling systems should be provided as determined by the program's needs.

Furnishings for the Laboratory/Classroom

A laboratory/classroom with movable tables and perimeter counters, sinks, and utilities provides maximum flexibility in the use of space. Fixed lab stations or service islands require more space, and sight lines may not be as good. If the room has fixed lab stations, students should enter the room primarily by the door in the classroom area.

If separate laboratories and classrooms are preferred and program and time requirements permit, two classrooms may share one laboratory. Specialized laboratories or spaces may be desired for advanced courses.

Sinks and Work Space. A large, deep sink with hot and cold water and 6 linear ft. (1.8 m) of adjacent counter space is needed for various purposes, such as cleaning large containers. Students' science investigations and cleanup require six sinks with water outlets along perimeter countertops or peninsulas. The sinks should be at least 18 in. (46 cm) in each dimension. Some means of acid dilution is necessary. Depending on the science program, EPA regulations may require an acid-neutralization system.

Each student will need at least 6 linear ft. (1.8 m) of horizontal work surface. Sturdy, movable two-student laboratory tables, at least 54 in. (137 cm) long, are needed. (Flat-topped desks or tablet arm chairs are sufficient for lecture space in a room with fixed lab stations.) Consider purchasing tables at least 80 in. (2 m) long to allow for the use of air tracks and other long devices.

If students will stand at their work stations, as in physical sciences laboratories, countertops 36 in. (91 cm) above the floor and tables 30 in. (76 cm) high are convenient for most high school students. Chairs of a variety of seat heights can accommodate students of various heights. A counter depth of 30 in. is desirable. Biological laboratories require seating capability, with tables preferably no more than 29 in. (74 cm) high, so that microscopes can be used comfortably. Ideally, all countertops and top surfaces in classrooms, preparation rooms, and storerooms will have chemical-resistant finishes. Chemical-resistant synthetic stone or epoxy resin is recommended for countertops and lab tabletops. Physics labs may need heavy-duty, wood-topped tables.

Storage and Wall Space. Provide base cabinets topped with counters along at least two walls for additional work and preparation space. All shelves and wall cabinets should be placed above base cabinets for safety reasons. Avoid particleboard assembly for casework unless a special sealant is used because it is prone to problems with water penetration.

Chalkboards or marker boards and tackable wall surfaces for maps and posters are also needed; allot space for apron storage, safety shower, and eye wash; and plan floor space for a demonstration table and equipment such as laboratory carts, animal or plant study center, and stream table.

For specialized storage, consider a storage closet (preferably at least 6 by 8 ft. or 1.8 by 2.4 m); shelves and cabinets of various sizes for science equipment such as skeletons, torsos, and microscopes; and at least 36 linear ft. (11 m) of bookshelves. Make sure enough lockable cabinets are available for teacher use and for storing student projects. Wide, shallow drawers are useful for storing posters. Tall cabinets can be used to store tote trays for individual students' supplies and kits.

Shelving at least 10 in. (25 cm) deep for books and 15 to 18 in. (38 to 46 cm) deep for equipment should be provided, mounted on standards that allow adjustment to different heights. In general, shelving 12 in. (30 cm) deep is preferred for chemistry needs, 18 in. (46 cm) for biological equipment, and 24 in. (61 cm) for physical science equipment.

Ceiling hooks (or for physics, steel pipes one inch in diameter, capable of supporting at least 100 lb. or 45 kg) are useful for hanging demonstration and experimental apparatus. Rails may be used under lab table tops to hold tote trays, books, and papers. It is important to provide storage for students' coats and book bags in the room or nearby to keep these items out of the way during lab work.

A pull-down audiovisual screen may be mounted at an angle in a front corner of the room, so that its use does not block the view of the chalkboard.

Lighting and Wiring. At least 10 covered, duplex electrical outlets with ground-fault-interrupter protection are needed at countertop height for the work spaces near the sinks, and additional outlets with ground-fault protection should be spaced frequently along the other walls for convenient use. For physical sciences laboratories, portable direct current sources protected by circuit breakers are also needed.

Provide wall outlets or recessed electrical floor boxes for computers and other equipment, as needed. Computers require separate circuits with surge protection and grounding. Also provide data connections to the school's computer network (or conduit and outlets for future connection) and wiring for voice, data, and video communications, as desired. A telephone or other provision for emergency help should be available in the science room or nearby.

The emergency shut-off controls for water, electrical service, and gas should be near the teacher's station, not far from the door, and not easily accessible to students. Master controls may include electric solenoid with key reactivation. The teacher's station should also have controls for room darkening and, if applicable, projection of video, computer imagery, and laserdisc media.

Lighting of a minimum of 50 foot-candles per square foot of floor surface (and 75 to 100 foot-candles at the work surface) is needed, as well as an emergency light (if adequate natural light is not available) in case of electricity failure. Room-darkening shades with edge tracks are necessary for various science activities. Dimming can be accomplished by using three-tube fluorescent lighting separately switched, so that one, two, or three tubes can be on at a time. Consider fluorescent lighting with parabolic lenses to reduce the glare on computer screens.

Equipment and Supplies

A source of heat, such as gas or hot plates, should be available for the students and teacher. Computers and appropriate software, television monitor, VCR, videodisc player, CD-ROM player, and an overhead projector (4,000 lumen for optimum use) are needed, as well as science materials and equipment including magnets, thermometers, hand lenses, microscopes, two-pan balances, measuring tools, models, plastic beakers, glass tubing, and containers. Laboratory carts with raised edges are required for the safe transport of chemicals. A projector for computer and video images is also recommended. Laptop computers with probeware, word-processing, and database software are desirable for use in field and outdoor studies as well as in the laboratory.

The following laboratory and field equipment is strongly recommended (as appropriate to the program):
- a clock with a sweep second hand
- overhead mirror above the demonstration table
- aquariums, terraria, and vivaria
- animal cages or study center
- incubator
- stereomicroscopes
- supply carts
- maps and globes

- astronomy equipment
- weather instruments
- stream table
- photography equipment
- microslide viewers
- videomicroscope
- tripod magnifiers
- germination/growth chamber
- refrigerator and dishwasher

Make arrangements to ensure adequate yearround care of animals, including backup heating and cooling systems for vacation periods.

Preparation and Storage Rooms

Ideally, a preparation room should be adjacent to the science room and near the storage areas. Consider providing a view window between the preparation and science rooms to facilitate supervision of students. Adequate ventilation and safety equipment are required. A large, deep sink with a rubber mat, hot and cold water, and 4 linear ft. (1.2 m) of adjacent counter space are needed for preparation for safety reasons. Also provide the following:
- two duplex electrical outlets with ground-fault protection
- gas cock or hot plate
- a variety of shelves and cabinets
- 9 linear ft. (2.7 m) of bookshelves

If a microwave oven is needed for demonstrations, it is best located in the preparation room. A refrigerator and dishwasher may be installed beneath a counter, and space may be allotted for other equipment (such as a distillation unit and autoclave), lab carts, and storage of glass tubing. A lowered section of countertop, 34 in. (86 cm) high, with knee space and drawers and with voice, data, and power outlets can serve as an office area for the teacher. Chemicals should not be stored in the preparation room, but should be kept in a separate, secure, and ventilated storage closet. State regulations must be observed.

Storage rooms supplement the storage areas in science rooms and provide needed security and specialized storage for large, expensive, or sensitive equipment.

A dedicated, locked chemical storage area that is well ventilated (12 air changes per hour forced-air ventilation) should be provided. Separation of incompatible materials is important, as are special precautions for flammable materials. Design shelves to inhibit spread of spills, and use wood or other corrosion-resistant materials for shelves and any attachments. It is desirable to limit shelf depth to 12 in. (30 cm) to prevent the storage of multiple bottles behind one another. Chemicals should not be stored in the same room with equipment or master shut-off controls because the chemicals could corrode metal parts.

Safety Considerations

For safety reasons, adequate circulation space and strict observation of the limit on class size are important. For students' protection, enclose any utility lines that are above or below service islands.

The science room should have splash-proof safety goggles for all students. Providing one sink with a soap dispenser for every four students also improves safety in the laboratory. It is a good idea to install a lip edge on any open shelves used for storing bottles or chemicals. Chemistry and some physical science rooms must have exhaust fume hoods.

In general, each laboratory and preparation room should have a centrally located
- safety shower (30 to 60 gallons per minute) and drain with a dual eyewash that provides a tempered flow of fresh aerated water for at least 15 minutes

- physician-approved first-aid kit
- fire blanket

For maximum safety, all laboratory, preparation, and storage rooms should have
- at least two locking entrances
- adequate ventilation
- a chemical spill control kit
- an ABC fire extinguisher at each exit
- smoke/heat detectors

Exhaust openings must be ventilated away from air intake openings. Fume hoods cannot replace the forced-air floor-to-ceiling ventilation necessary in science rooms.

Separate student project rooms also require appropriate safety equipment and egress.

Finally, be sure to provide supervision for custodians who clean chemical storage areas unless they have had specific training.

Earthquake Precautions

Recommendations and requirements for earthquake and hurricane precautions for school facilities can be obtained from the state disaster department.

Because books and equipment will fall off open shelves during earthquakes, cabinets with hinged doors and positive latches are recommended in areas subject to seismic activity. These cabinets should be bolted to walls and partitions. Deep tracks will help prevent sliding cabinet doors from being jolted out of their tracks by the upward motion of an earthquake. Lips or rods can help keep items from sliding off shelves; ideally, shelving for chemical storage will have individual recesses for each container.

It is advisable to clamp or bolt equipment to counters wherever possible. If computers are mounted on carts, the carts may be kept in cabinets when not in use. Major building codes specify requirements for walls, ceilings, and equipment.

Adaptations for Students with Disabilities

To accommodate students with physical disabilities, additional space is needed for specialized equipment, such as wheelchairs, and for aides who may accompany the student. An appropriate work area requires approximately twice the usual amount of space—that is, the equivalent of two work stations per student. Specialized equipment and furniture are also needed.

If each room has sufficient space, portable equipment and adjustable furniture can be used to make at least one work station in every room accessible to students with disabilities. Provision of adequate space requires significantly less expense than the installation of permanent equipment, which often goes unused.

Ways to ensure accessibility include
- providing Braille equivalents on labels for switches and equipment
- using wire pulls on cabinets and lever-controlled faucets on sinks
- equipping emergency eyewashes with extendable hoses

All wall-mounted objects should be above base cabinets so blind students won't risk bumping into them.

The *ADA Accessibility Guidelines for Buildings and Facilities* (ADAAG) defines a number of requirements for maximum counter heights, reach ranges, and grasping and twisting limitations. For example, the ADAAG specifies that sinks used by disabled students must be no more than 6.5 in. (17 cm) deep and must be mounted to accommodate students in wheelchairs. (For adults, this would mean a counter no more than 34 in. or 86 cm, high, with 27 in. or 69 cm, minimum vertical knee space.)

The school district advisor should have information about local, state, and federal guidelines. The ADA, enforced through the United States Department of Justice, requires compliance with ADAAG or *Uniform Federal Accessibility Standards* (UFAS) requirements.

Other Teaching Environments

Technology labs can offer computer simulation technology, which is used in both science and technology programs. A seminar room equipped with a computer work station and electronic communications system for electronic presentations is useful for small-group work and can be shared with other departments. A planetarium is recommended for the school district.

An optimal science program includes access to natural settings in which to study basic science concepts. (Be sure to include the outdoor areas for science activities in the initial facilities design plan.) The following are all desirable:
- native plants on the school grounds
- a garden
- greenhouse with water source
- storage shed for garden tools
- nature trails
- natural outdoor areas with a variety of environments
- educational kiosks

A less expensive alternative to a greenhouse is a small (4 ft. deep by 8 ft. wide, or 1.2 by 2.4 m) lean-to structure protruding from the wall of a science room, glazed with insulating glass, and equipped with supplementary day and night lighting and a floor drain.

"Designing High School Science Facilities" was written by James T. Biehle, Inside/Out Architecture, Inc., Clayton, Missouri; Sandra S. West, Department of Biology, Southwest Texas State University, San Marcos; and LaMoine L. Motz, Oakland County Schools, Waterford, Michigan.

Other contributors were Patricia S. Bowers, Center for Mathematics and Science, University of North Carolina, Chapel Hill; Terry Kwan, TK Associates, Brookline, Massachusetts; and Victor M. Showalter, Purdue University, West Lafayette, Indiana.

Resources for the Road

ADA Accessibility Guidelines for Buildings and Facilities. (1991, July 26). *Federal Register, 56* (144).

American Association for the Advancement of Science. (1991). *Barrier Free in Brief: Laboratories and Classrooms in Science and Engineering.* Washington, DC: Author.

American Chemical Society. (1995). *Safety in Academic Chemistry Laboratories* (6th ed.). Washington, DC: Author.

Biehle, James T. (1995, November). Six Science Labs for the 21st Century. *School Planning and Management, 34* (9), 39–42.

Biehle, James T. (1995, May). Complying with Science. *American School and University, 67* (9), 54–56. (Discusses ADA issues in science labs.)

California Department of Education, Science and Environmental Education Unit. (1993). *Science Facilities Design for California Public Schools.* Sacramento, CA: Author.

Flinn Biological Catalog/Reference Manual. (1996). Batavia, IL: Flinn Scientific, Inc. (P.O. Box 219, Batavia, IL 60510; contains advice on safety in the laboratory.)

Flinn Chemical Catalog/Reference Manual. (1996). Batavia, IL: Flinn Scientific, Inc. (Contains advice on safety in the laboratory.)

Florida Department of Education. (1993). *Science for All Students: The Florida Pre K–12 Science Curriculum Framework.* Tallahassee, FL: Author.

Florida Department of Education. (1992). *Science Safety: No Game of Chance!* Tallahassee, FL: Author.

Governor's Committee on High School Science Laboratories for the 21st Century. (1992). *Look of the Future: Report of the Governor's Committee on High School Science Laboratories for the 21st Century.* Baltimore, MD: State of Maryland, Public School Construction Program.

Kaufman, James A. *The Kaufman Letter* (quarterly newsletter on safety issues). Natick, MA: James A. Kaufman and Associates. (192 Worcester Road, Natick, MA 01760).

Los Angeles, Orange, and San Diego County Offices of Education. (1989). *Remodeling and Building Science Instruction Facilities in Elementary, Middle, Junior, and Senior High Schools.* (1989). Downey, CA: Los Angeles County Office of Education. (Also available from Orange County Department of Education, Costa Mesa, CA, and San Diego County Office of Education, San Diego, CA.)

Madrazo, Gerry M., and Motz, LaMoine L. (Eds.). (1993). *Fourth Sourcebook for Science Supervisors.* Arlington, VA: National Science Teachers Association.

Mione, Lawrence V. (1995). *Facilities Standards for Technology in New Jersey Schools.* Trenton, NJ: New Jersey State Department of Education.

National Science Teachers Association. (1995). Laboratory Science (Position statement). In *NSTA Handbook 1995–1996* (pp. 209–212). Arlington, VA: Author.

National Science Teachers Association. (1993). *Science in the Elementary Science Classroom* (Rev. ed.). Arlington, VA: Author.

Public Schools of North Carolina. (1991). *North Carolina Public Schools Facility Standards: A Guide for Planning School Facilities.* Raleigh, NC: North Carolina Department of Public Instruction, School Planning.

Public Schools of North Carolina. (1992). *Furnishing and Equipment Standards: A Guide for Planning and Equipping New Facilities and Evaluating Existing Schools.* Raleigh, NC: North Carolina Department of Public Instruction, School Planning.

Reese, Kenneth M. (Ed.). (1985). *Teaching Chemistry to Physically Handicapped Students* (Rev. ed.). Washington, DC: American Chemical Society.

continued next page

School Facilities Branch, Maryland State Department of Education. (1994). *Science Facilities Design Guidelines.* Baltimore, MD: Author.

Showalter, Victor M. (Ed.). (1984). *Conditions for Good Science Teaching.* Arlington, VA: National Science Teachers Association.

Six Secrets To Holding a Good Meeting Every Time [Brochure]. (n.d.). Saint Paul, MN: 3M Company, Audiovisual Division.

Texas Education Agency (TEA). (1989). *Planning a Safe and Effective Science Learning Environment.* Austin, TX: Author. (Available from Publications, Distribution, and Fees Office, TEA, 1701 North Congress Avenue, Austin, TX 78701-1494.)

19 Texas Administrative Code, Chapter 61, Subchapter H (School Facilities Standards), § 61.102. (Available from Director, School Facilities, TEA, 1701 North Congress Avenue, Austin, TX 78701-1494.)

Wang, Denis. (1994, February). A Working Laboratory. *The Science Teacher, 61* (2), 26-29.

Ward, John. (1992, September). Shopping for Science. *The Science Teacher, 59* (6), 28–33.

Ward, Susan, and West, Sandra S. (1990, May). Accidents in Texas High School Chemistry Labs. *The Texas Science Teacher, 19* (2), 14–19.

West, Sandra. S. (1991, September). Lab Safety. *The Science Teacher, 58* (6), 45–49.

Young, J. R. (1972). A Second Survey of Safety in Illinois High School Laboratories. *Journal of Chemical Education, 49* (1), 55. (Contains research on space necessary for science safety in the laboratory.)

Appendix D
Resources for the Road: The CD-ROM Collection

The optional compact disc that accompanies *Pathways to the Science Standards* contains a majority of the articles cited in the "Resources for the Road" sections of this book. This includes all the articles from the NSTA journals *The Science Teacher, Science Scope,* and *Science and Children* as well as selected items from other sources.

You can search a specific article on the disc by using the author's last name or the *first* word of the title of the article.

We hope that the wealth of material in these resources will prove invaluable to you as you follow your own pathway to putting the vision of the National Science Education Standards into action.

DATE DUE

GAYLORD   PRINTED IN U.S.A.